FOUNDATIONS OF INFINITESIMAL CALCULUS

H. JEROME KEISLER
University of Wisconsin

Prindle, Weber & Schmidt, Incorporated
20 Newbury Street, Boston, Massachusetts

ISBN 87150-215-1

Printed in the United States of America.

Second printing: December 1977

PREFACE

In 1960 Abraham Robinson solved the three hundred year old problem of giving a rigorous development of the calculus based on infinitesimals. Robinson's achievement may be one of the major mathematical advances of the century. This is an exposition of Robinson's infinitesimal calculus at the advanced undergraduate level. It is entirely self-contained but is keyed to the first year college text Elementary Calculus by H. Jerome Keisler, published in 1976. This monograph can be used as a quick introduction to the subject for mathematicians, as background material for instructors using the book Elementary Calculus, or as a text for an undergraduate seminar.

The basic concepts of the calculus were originally developed in the seventeenth and eighteenth centuries using the intuitive notion of an infinitesimal. When the calculus was put on a rigorous basis in the nineteenth century, infinitesimals were rejected in favor of the ε, δ approach, because mathematicians had not yet discovered a correct treatment of infinitesimals. Since then generations of students have been taught that infinitesimals do not exist and should be avoided.

The actual situation, as suggested by Leibniz and carried out by Robinson, is that one can form the hyperreal number system by adding infinitesimals to the real number system, and obtain a powerful new tool in analysis. The reason Robinson's discovery did not come sooner is that the axioms needed to describe the hyperreal numbers are of a kind which were unfamiliar to mathematicians until recently. Robinson used methods from the branch of mathematical logic called model theory which developed in the 1950's.

Robinson called his method nonstandard analysis because it uses a nonstandard model of analysis. The older name infinitesimal analysis is perhaps more appropriate. The method is surprisingly adaptable and is likely to influence many areas of pure and applied mathematics. It is beginning to be used in such fields as economics and physics as a source of new mathematical models.

The purpose of this monograph and of Elementary Calculus is to make infinitesimals available to mathematicians and students. The infinitesimals provided the intuition for the original development of the calculus and should help students as they repeat this development. The book Elementary Calculus treats infinitesimal calculus at the simplest possible level, and gives plausibility arguments instead of proofs of theorems whenever it is appropriate. This monograph presents the subject from a more advanced viewpoint and includes proofs of almost all of the theorems stated in Elementary Calculus.

The chapter numbers in this monograph match the chapter numbers in Elementary Calculus, and after each section heading the corresponding sections of Elementary Calculus are indicated in parentheses. In Chapter 1 the hyperreal numbers are first introduced with a set of axioms and their algebraic structure is studied. Then in Section 1E the hyperreal numbers are constructed in a simple way from the real numbers. Chapters 2 through 12 contain a rigorous development of infinitesimal calculus based on the axioms in Chapter 1. The only prerequisites are the traditional three semesters of calculus and a certain amount of mathematical maturity. In particular, the material is presented without using notions from mathematical logic. We will use some elementary set-theoretic notation familiar to all mathematicians, for example the function concept and the symbols $\emptyset$, $A \cup B$, $\{x \in A: P(x)\}$.

An optional chapter between Chapters 1 and 2, Chapter 1*, provides a bridge between the simple treatment of infinitesimal calculus in this monograph and the more advanced subject of infinitesimal analysis found in the research literature. To go beyond infinitesimal calculus one should at least be familiar with some basic notions from logic and model theory. Chapter 1* explains the use of mathematical logic, superstructures, and internal and external sets, and presents a more sophisticated construction of the hyperreal numbers using ultrapowers.

With one exception the results in this exposition can be found in the existing literature in infinitesimal analysis (especially the books of Robinson and Stroyan) and the traditional advanced calculus. The exception is the simple set of axioms given here (and in Elementary Calculus) and its equivalence to Robinson's approach. These axioms make it possible to present infinitesimal calculus at the college freshman level, avoiding concepts from mathematical logic. Frequently standard results are given alternate proofs using infinitesimals. In some cases a standard result which is beyond the scope of beginning calculus is rephrased as a simpler infinitesimal result and used effectively in Elementary Calculus; some examples are the Infinite Sum Theorem, and the two-variable criterion for a global maximum.

For further reading in infinitesimal analysis and model theory, see the following books.

C. C. Chang and H. J. Keisler. Model Theory. Amsterdam, London: North-Holland Publ. Co., 1973.

A. Robinson. Non-standard Analysis. Amsterdam, London: North-Holland Publ. Co., 1970.

K. Stroyan with the collaboration of W. A J. Luxemburg. Introduction to the Theory of Infinitesimals. New York: Academic Press, 1976.

Two good texts in ordinary advanced calculus are:

R. C. Buck with the collaboration of Ellen F. Buck. Advanced Calculus, Second Edition. New York: McGraw Hill, 1956.

W. Rudin. Principles of Mathematical Analysis. New York: McGraw Hill, 1964.

The following article contains 275 references in infinitesimal analysis.

D. Randolph Johnson. Bibliography of Nonstandard Analysis - June, 1975. Lecture Notes in Mathematics 3, Department of Mathematics, University of Pittsburgh, Pittsburgh, Pa. (Mimeographed).

I wish to thank Jon Barwise of the University of Wisconsin and the University of California at Los Angeles, Donald McKillican of the University of Saskatchewan, and Keith Stroyan of the University of Iowa for valuable advice in preparing this monograph. I also thank Sue Balsiger for her excellent typing of the manuscript.

H. Jerome Keisler

CONTENTS

[†]Corresponding sections of <u>Elementary Calculus</u> are indicated in parentheses.

1 THE HYPERREAL NUMBERS

We shall assume that the reader is familiar with the real number system and develop a new object, the hyperreal number system. The definition of the real numbers and the basic existence and uniqueness theorems are briefly outlined in an Appendix, Section 1F, at the end of this chapter. This Appendix also explains some useful notions from modern algebra, such as a ring, a complete ordered field, an ideal, and a homomorphism. If any of these terms are unfamiliar, you should read through the Appendix. We do not require any knowledge of modern algebra except for modest vocabulary.

In Sections 1A-1D we introduce axioms for the hyperreal numbers and obtain some first consequences of the axioms. Then in Section 1E we give a simple construction of the hyperreal numbers, that is, we prove that there exists a structure which satisfies the axioms.

1A STRUCTURE OF THE HYPERREAL NUMBERS (§1.5, §1.6)

In this section we assume only that

AXIOM A

R is a complete ordered field.

AXIOM B

R^* is a proper ordered field extension of R.

We call R the field of real numbers and R^* the field of hyperreal numbers. In the next section we shall introduce two powerful additional axioms which are needed for the calculus. However, the algebraic facts about infinitesimals which underlie the intuitive picture of the hyperreal line follow from Axioms A and B alone.

DEFINITION

An element $x \in R^*$ is infinitesimal if $|x| < r$ for all positive real r; finite if $|x| < r$ for some real r; infinite if $|x| > r$ for all real r.

Two elements $x, y \in R^*$ are said to be infinitely close, $x \approx y$, if $x - y$ is infinitesimal. (Thus x is infinitesimal if and only if $x \approx 0$.)

Notice that the only real infinitesimal is 0. We shall prove later that there exist nonzero infinitesimals in R^*.

DEFINITION

Given a hyperreal number $x \in R^*$, the monad of x is the set

$$\text{monad}(x) = \{y \in R^*: x \approx y\}.$$

The galaxy of x is the set

$$\text{galaxy}(x) = \{y \in R^*: x - y \text{ is finite}\}.$$

Thus monad(0) is the set of infinitesimals and galaxy(0) is the set of finite hyperreal numbers. In Elementary Calculus the pictorial device of an infinitesimal microscope is used to illustrate part of a monad, and an infinite telescope is used to illustrate a part of an infinite galaxy. Figure 1 shows how the hyperreal line R^* is drawn. In Section 2B we shall give a rigorous treatment of infinitesimal microscopes and telescopes so the instructor can use them in new situations.

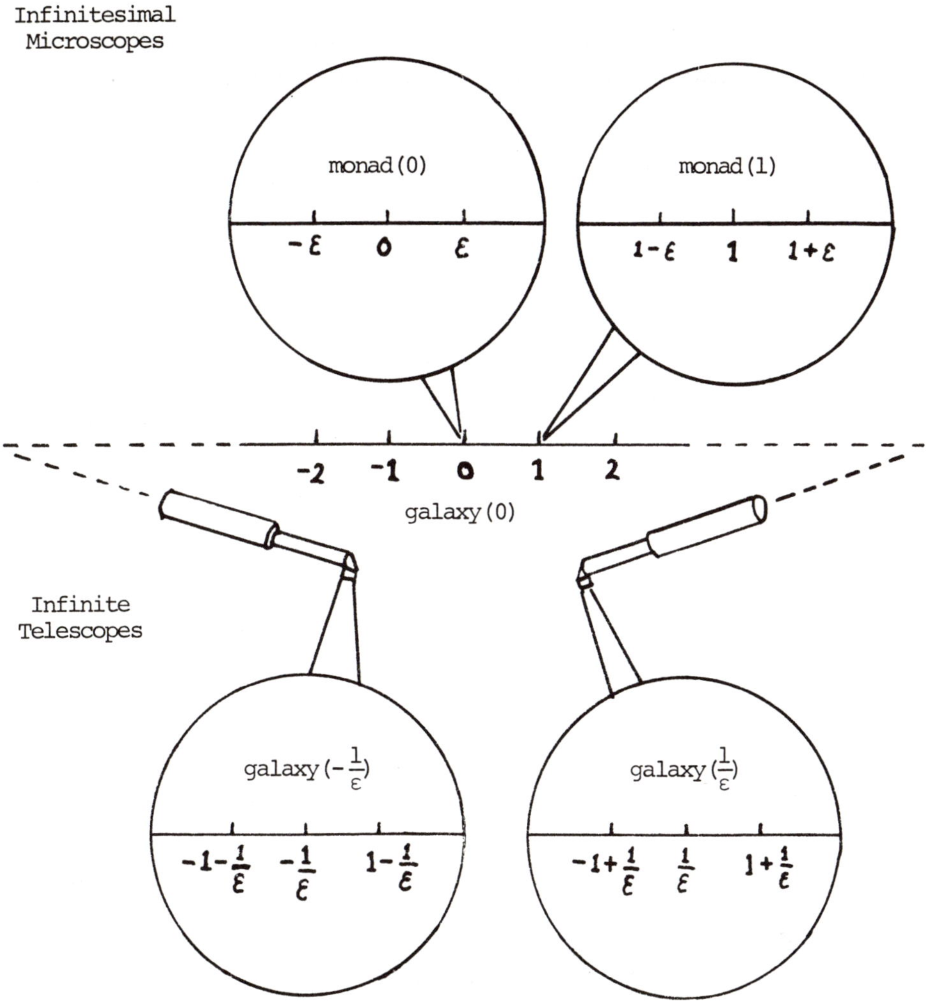

The Hyperreal Line

Figure 1

We now describe the algebraic structure of R^*.

THEOREM 1

The set galaxy(0) of finite elements is a subring of R^*. That is, sums, differences, and products of finite elements are finite.

Proof Suppose x and y are finite, say

$$|x| < r, \ |y| < s$$

where r and s are real. Then

$$|x + y| < r + s, \ |x - y| < r + s, \ |xy| < rs,$$

so x + y, x - y, and xy are finite. ⊣

COROLLARY 2 Any two galaxies are either equal or disjoint.

Proof For each $x \in R^*$, the galaxy of x is the coset of x modulo galaxy(0),

$$\text{galaxy}(x) = \{x + a: a \in \text{galaxy}(0)\}.$$ ⊣

THEOREM 3

The set monad(0) of infinitesimal elements is a subring of R^* and an ideal in galaxy(0). That is:

(i) Sums, differences, and products of infinitesimals are infinitesimal.

(ii) The product of an infinitesimal and a finite element is infinitesimal.

Proof Let $\varepsilon, \delta \approx 0$. For each positive real r, $|\varepsilon| < r/2$, $|\delta| < r/2$, whence $|\varepsilon + \delta| < r$, $|\varepsilon - \delta| < r$. Hence $\varepsilon + \delta$ and $\varepsilon - \delta$ are infinitesimal.

Let b be finite, say $|b| < t$, $t \in R$. Then for any positive real r we have $|\varepsilon| < r/t$, $|b\varepsilon| < (r/t)t = r$. Therefore εb is infinitesimal. ⊣

COROLLARY 4 Any two monads are equal or disjoint. The relation $x \approx y$ is an equivalence relation on R*.

Proof For each $x \in R^*$, monad(x) is the coset of x modulo monad(0),

$$\text{monad}(x) = \{x + \varepsilon : \varepsilon \in \text{monad}(0)\}.$$

From the definition of monad and $x \approx y$ we see that

$$x \approx y \text{ if and only if } \text{monad}(x) = \text{monad}(y),$$

so $x \approx y$ is an equivalence relation. ⊣

THEOREM 5

(i) x is infinite if and only if x^{-1} is infinitesimal.

(ii) monad(0) is a maximal ideal in galaxy(0). That is, there is no ideal I in galaxy(0) such that $\text{monad}(0) \subsetneq I \subsetneq \text{galaxy}(0)$.

Proof (i) The following are equivalent:

$|x| \geq r$ for all positive real r.

$|x^{-1}| \leq r^{-1}$ for all positive real r.

x^{-1} is infinitesimal.

(ii) Let I be an ideal containing monad(0) and let $b \in I - \text{monad}(0)$. By (i), b^{-1} is not infinite since $b = (b^{-1})^{-1}$ is not infinitesimal. Then $b^{-1} \in \text{galaxy}(0)$, so $1 = b\cdot(b^{-1}) \in I$. Then for any $c \in \text{galaxy}(0)$, $1\cdot c = c \in I$, so $I = \text{galaxy}(0)$. ⊣

THEOREM 6 (Standard Part Theorem)

Every finite $x \in R^*$ is infinitely close to a unique real number $r \in R$. That is, every finite monad contains a unique real number.

Proof Let $x \in R^*$ be finite.

Uniqueness: Suppose r and s are real and $r \approx x$, $s \approx x$. Since $\approx$ is an equivalence relation we have $r \approx s$, whence $r - s \approx 0$. But $r - s$ is real, so $r - s = 0$ and $r = s$.

Existence: Let $X = \{s \in R : s < x\}$. X is nonempty. X has an upper bound because there is a positive real number r such that $|x| < r$, whence $-r < x < r$, so $-r \in X$ and r is an upper bound of X.

Let t be the least upper bound of X. For every positive real r we have

$$x \leq t + r, \; x - t \leq r$$

and

$$t - r \leq x, \; -(x - t) \leq r.$$

It follows that $x - t \approx 0$, so $x \approx t$. ⊣

DEFINITION

Given a finite $x \in R^*$, the unique real $r \approx x$ is called the <u>standard part</u> of x, in symbols $r = st(x)$. If x is infinite, st(x) is undefined.

COROLLARY 7 Let x and y be finite.

(i) $x \approx y$ if and only if $st(x) = st(y)$.
(ii) $x \approx st(x)$.
(iii) If $r \in R$ then $st(r) = r$.
(iv) If $x \leq y$ then $st(x) \leq st(y)$.

<u>Proof of (iv)</u> We have

$$x = st(x) + \varepsilon, \; y = st(y) + \delta$$

for infinitesimal ε and δ. Assume $x \leq y$. Then

$$st(x) + \varepsilon \leq st(y) + \delta,$$

$$st(x) \leq st(y) + (\delta - \varepsilon).$$

For any positive real r,

$$\delta - \varepsilon < r,$$

$$st(x) < st(y) + r,$$

and therefore

$$st(x) \leq st(y).$$ ⊣

THEOREM 8

(i) There exist positive and negative infinitesimals in R*.
(ii) R* has positive and negative infinite elements.

<u>Proof</u> Since R* is a proper extension of R there is at least one element $x \in R^* - R$. If x is finite then $|x - st(x)|$ is a positive infinitesimal. If x is infinite then $|1/x|$ is a positive infinitesimal by Theorem 5. Thus there exists a positive infinitesimal ε. The elements $-\varepsilon$, $1/\varepsilon$, and $-1/\varepsilon$ are respectively negative infinitesimal, positive infinite, and negative infinite. ⊣

Actually, one can go on to show that there are infinitely many infinitesimals and infinitely many galaxies in R*. Moreover, each galaxy is partitioned into infinitely many monads. Each monad has a complicated structure; for example, the mapping $H \to x + \frac{1}{H}$ maps the infinite elements of R* one-one onto monad(x) − {x}.

We caution the reader that monad(0) is an ideal in galaxy(0) but is only a subring, not an ideal, in R*. In other words, the product of an infinitesimal and an infinite element need not be infinitesimal. In fact, given a positive infinitesimal ε, we see that

$$\varepsilon^2 \cdot \frac{1}{\varepsilon} \text{ is infinitesimal,}$$

$$\varepsilon \cdot \frac{1}{\varepsilon} \text{ is finite but not infinitesimal,}$$

$$\varepsilon \cdot \frac{1}{\varepsilon^2} \text{ is infinite.}$$

This corresponds to the intuitive principle that $0 \cdot \infty$ is an "indeterminate form."

We conclude with an important result about standard parts.

THEOREM 9

The standard part function is a homomorphism of the ring galaxy(0) onto the field of real numbers.

That is, for finite x and y,

(i) $st(x + y) = st(x) + st(y)$,

(ii) $st(x - y) = st(x) - st(y)$,

(iii) $st(xy) = st(x)st(y)$.

Proof Let $x = r + \varepsilon$, $y = s + \delta$ where $r = st(x)$, $s = st(y)$. Then ε and δ are infinitesimal.

(i) $st(x + y) = st((r + \varepsilon) + (s + \delta))$
$= st((r + s) + (\varepsilon + \delta))$
$= r + s.$

(ii) is similar to (i).

(iii) $st(xy) = st((r + \varepsilon)(s + \delta))$
$= st(rs + r\delta + s\varepsilon + \varepsilon\delta)$
$= rs,$
since $r\delta + s\varepsilon + \varepsilon\delta$ is infinitesimal.

Thus st is a homomorphism of galaxy(0) into R. It is obviously onto R, because for $r \in R$, $st(r) = r$. ⊣

COROLLARY 10 Let x and y be finite.

(i) If $st(y) \neq 0$ then $st(\frac{x}{y}) = \frac{st(x)}{st(y)}$.

(ii) If $y = \sqrt[n]{x}$ then $st(y) = \sqrt[n]{st(x)}$.

Proof (i) $st(x) = st(\frac{x}{y} \cdot y) = st(\frac{x}{y}) \cdot st(y)$ and (i) follows.

(ii) The proof of (ii) is similar to (i). ⊣

1B AXIOMS FOR THE HYPERREAL NUMBERS (§1.7-1.9)

The first two axioms were given in Section 1A. Before giving a precise statement of the remaining two axioms for the hyperreal numbers, we describe the intuitive idea. The real and hyperreal numbers are related by a mapping * such that:

(1) With each relation X on R there is associated a relation X* on R*, called the natural extension of X.

(2) The relations on R have the same "elementary properties" as their natural extensions on R*.

The difficulty in making (2) precise is that we must explain exactly what an "elementary property" is. The properties "$X \subseteq Y$", "X is a function", and "X is a binary function which satisfies the Commutative Law" are elementary. On the other hand, the Archimedean Property and the Completeness Property must not be elementary, because no proper extension of R is an Archimedean or complete ordered field. In most expositions of the subject an "elementary property" is taken to be any sentence in first order logic. However, it is not appropriate to begin a calculus course by explaining first order logic to the students because they have not yet been exposed to the right sort of examples. It is better to learn calculus first, and at some later time use the ε,δ conditions from calculus as meaningful examples of sentences in first order logic. Fortunately, the notion of a sentence in first order logic is not necessary at all in stating the axioms. It turns out that a simpler concept which is within the experience of beginning students is sufficient. This is the concept of a (finite) system of equations and inequalities. We shall see in Chapter 1* that we get equivalent sets of axioms using either the simple concept of a system of equations and inequalities or the more sophisticated concept of a sentence of first order logic.

The main objects of study in elementary calculus are partial functions of n real variables. Our plan is to have an axiom corresponding to (1) but for partial functions instead of relations, and then an axiom corresponding to (2) where an "elementary property" means a system of equations and inequalities in these functions.

In the next few paragraphs we shall explain exactly what is meant by a system of equations and inequalities.

R^n denotes the set of all n-tuples of elements of R. A <u>real function</u> of n variables is a mapping f from a subset of R^n into R.

The letters x,y,z,... are called <u>variables</u>. We think of them as varying over the set R of real numbers. A real number $c \in R$ is also called a <u>constant</u>. Expressions built up from variables and constants by applying real functions are called terms. For example,

$$x, \quad c, \quad x + c, \quad f(x), \quad g(c,x,f(y))$$

are terms. Here is a precise definition of a term.

A <u>term</u> is an expression which can be built up using the following rules:

(T1) Every variable is a term.

(T2) Every real constant is a term.

(T3) If $\tau_1,\ldots,\tau_n$ are terms and f is a real function of n variables, then $f(\tau_1,\ldots,\tau_n)$ is a term.

A term which contains no variables is called a <u>constant term</u>. A constant term is either undefined or has value equal to some real number. In particular, f(c) is undefined if c is not in the domain of f, and has value d if f(c) = d. The value of a constant term is computed step by step; thus the value of

$$g[f(5) + 3, f(f(2))]$$

is computed by first computing f(5), then f(5) + 3, then f(2), then f(f(2)), and then g[f(5) + 3, f(f(2))]. If at any stage in the computation we reach an undefined part, the whole constant term is considered to be undefined. For example, since $\sqrt{-6}$ is undefined in the reals, the constant terms $\sqrt{-6} - \sqrt{-6}$ and $\sqrt{-6}/\sqrt{-6}$ are also undefined.

By an <u>equation</u> we mean an expression $\sigma = \tau$ where σ and τ are terms. By an <u>inequality</u> we mean an expression of one of the forms

$$\sigma \leq \tau, \ \sigma < \tau, \ \sigma \neq \tau$$

where σ and τ are terms. By a <u>formula</u> we mean an equation or inequality between two terms. A <u>system of formulas</u> is a finite set of formulas.

The last notion we need is that of a solution. Consider a system S of formulas which contains the variables $x_1,\ldots,x_n$. By a <u>solution</u> of S we mean an n-tuple $(c_1,\ldots,c_n)$ of constants such that when the x_i are replaced by c_i in S, each term in S is defined and each formula in S is true.

Notice that the set of solutions of the trivial equation

$$f(x) = f(x)$$

is the domain of f rather than the set of all reals. If f(c) is undefined then c is not a solution of the equation f(x) = f(x).

The notion of a system of formulas is easily motivated in an elementary calculus course, because the first step in solving a "story problem" is to set the problem up as a system of formulas.

The following axioms describe a triple (R,R*,*), where R is called the field of <u>real numbers</u>, R* the field of <u>hyperreal numbers</u>, and * the <u>natural extension mapping</u>.

AXIOM A

R is a complete ordered field.

AXIOM B

R* is a proper ordered field extension of R.

AXIOM C (Function Axiom)

For each real function f of n variables there is a corresponding hyperreal function f* of n variables, called the natural extension of f. The field operations of R* are the natural extensions of the field operations of R.

AXIOM D (Solution Axiom)

If two systems of formulas have exactly the same real solutions, they have exactly the same hyperreal solutions.

By a hyperreal solution of a system of formulas S with the variables $x_1,\ldots,x_n$ we mean an n-tuple $(c_1,\ldots,c_n)$ of hyperreal numbers such that all the formulas in S are true when each function is starred and each x_i is replaced by c_i.

The Solution Axiom can be motivated in the following way. Often a real function f is defined by a rule of the form:

$$f(x_1,\ldots,x_n) = y \text{ if and only if } S$$

where S is a system of formulas. The Solution Axiom has the following special case:

If a real function f is defined by the rule

$$f(x_1,\ldots,x_n) = y \text{ if and only if } S,$$

then the natural extension f* is defined by the same rule applied to the hyperreal numbers.

For example, the square root function on the reals is defined by the rule

$$\sqrt{x} = y \text{ if and only if } \{y^2 = x,\ 0 \le y\}.$$

By the Solution Axiom, the natural extension $\sqrt{\ }^*$ is defined by the same rule where x and y vary over the hyperreal numbers.

1C CONSEQUENCES OF THE SOLUTION AXIOM (§1.9)

The Solution Axiom is much more powerful than it looks. We shall postpone a full explanation of its scope to the advanced Chapter 1*, because this explanation requires notions from logic. At this point we shall derive some easy but general consequences of the Solution Axiom which will facilitate our development of the calculus. We shall then prove that Axioms A-D above are equivalent to the slightly different axioms given in the textbook Elementary Calculus.

THEOREM 1

(i) If a system of formulas S is true for all real numbers, it is true for all hyperreal numbers.

(ii) If a system of formulas S has no real solutions, it has no hyperreal solutions.

Proof (i) Since S is true for all real numbers, it has the same real solutions as the equation $x = x$. By the Solution Axiom, S has the same hyperreal solutions as the equation $x = x$, so S is true for all hyperreal numbers.

(ii) The proof is similar but uses the inequality $x \neq x$ instead of the equation $x = x$. ⊣

THEOREM 2

Let f be a real function of n variables.

(i) If $f(c_1,\ldots,c_n)$ is defined then $f^*(c_1,\ldots,c_n) = f(c_1,\ldots,c_n)$.

(ii) If $f(c_1,\ldots,c_n)$ is undefined then $f^*(c_1,\ldots,c_n)$ is undefined.

Proof (i) Let $f(c_1,\ldots,c_n) = c$. The system of formulas

$$f(c_1,\ldots,c_n) = c,\ x = x$$

is true for all real numbers, hence for all hyperreal numbers. Therefore

$$f^*(c_1,\ldots,c_n) = c.$$

(ii) Suppose $f(c_1,\ldots,c_n)$ is undefined. Then the system of formulas

$$f(c_1,\ldots,c_n) = f(c_1,\ldots,c_n),\ x = x$$

has no real solutions. Thus it has no hyperreal solutions, so $f^*(c_1,\ldots,c_n)$ is undefined. ⊣

From now on we shall drop the asterisk on the natural extension of f and write $f(c_1,\ldots,c_n)$ for the value of f* at $(c_1,\ldots,c_n)$, whether the c_i's are real or hyperreal. This will cause no trouble because, by Theorem 2, $f(c_1,\ldots,c_n)$ and $f^*(c_1,\ldots,c_n)$ are equal when the c_i's are real.

EXAMPLE The standard part function st is an example of a hyperreal function which is not the natural extension of any real function. The domain of st is the set of all finite hyperreal numbers. st cannot be a natural extension because the equation $st(x) = x$ holds for all real numbers but not for all hyperreal numbers.

The standard part function is the only hyperreal function we shall use which is not the natural extension of a real function.

It is useful to be able to treat certain statements about functions or relations as formulas when applying the Solution Axiom. We shall say that a statement P is <u>equivalent</u> to a formula F if P has exactly the same real and hyperreal solutions as F.

THEOREM 3

Let f be a real function of n variables. Each of the following statements is equivalent to an equation.

(i) $f(x_1,\ldots,x_n)$ is defined.

(ii) $f(x_1,\ldots,x_n)$ is undefined.

Proof (i) By our definition of a solution, the statement (i) is equivalent to the equation

$$f(x_1,\ldots,x_n) = f(x_1,\ldots,x_n).$$

(ii) Let g be the characteristic function of the domain of f,

$$g(x_1,\ldots,x_n) = \begin{cases} 1 & \text{if } f(x_1,\ldots,x_n) \text{ is defined,} \\ 0 & \text{if } f(x_1,\ldots,x_n) \text{ is undefined.} \end{cases}$$

We show that the statement (ii) is equivalent to the equation

$$g(x_1,\ldots,x_n) = 0.$$

We first note that the equation

$$g(x_1,\ldots,x_n) = g(x_1,\ldots,x_n)^2$$

holds for all real numbers and hence for all hyperreal numbers. Therefore $g(x_1,\ldots,x_n)$ is defined and has value 0 or 1 for all hyperreal numbers. Moreover, the equations

$$g(x_1,\ldots,x_n) = 1, \tag{1}$$

$$f(x_1,\ldots,x_n) = f(x_1,\ldots,x_n) \tag{2}$$

have the same real and hence hyperreal solutions. Therefore the following are equivalent for all real or hyperreal numbers $c_1,\ldots,c_n$.

$f(c_1,\ldots,c_n)$ is undefined

$(c_1,\ldots,c_n)$ is not a solution of (2)

$(c_1,\ldots,c_n)$ is not a solution of (1)

$g(c_1,\ldots,c_n) = 0$. ⊣

THEOREM 4

Suppose S and T are systems of formulas, and every real solution of S is a real solution of T. Then every hyperreal solution of S is a hyperreal solution of T.

Proof S and S ∪ T have the same real solutions. Hence S and S ∪ T have the same hyperreal solutions, so every hyperreal solution of S is a hyperreal solution of T. ⊣

Our next theorem extends Theorem 4 to the case where the system of formulas T has more variables than S.

DEFINITION

Let T be a system of formulas with variables $x_1,\dots,x_k,\dots,x_n$. A partial real solution of T is a k-tuple $(c_1,\dots,c_k)$ of reals which can be extended to a real solution $(c_1,\dots,c_k,\dots,c_n)$ of T. A partial hyperreal solution is defined similarly.

THEOREM 5 (Partial Solution Theorem)

Let S be a system of formulas with the variables $x_1,\dots,x_k$ and T a system of formulas with the variables $x_1,\dots,x_k,\dots,x_n$. The following are equivalent.

(i) Every real solution of S is a partial real solution of T.
(ii) Every real solution of S is a partial hyperreal solution of T.
(iii) Every hyperreal solution of S is a partial hyperreal solution of T.

Proof We prove (i) → (iii) → (ii) → (i). To simplify notation let S have the single variable x and T the two variables x, y.

Assume (i). For each real solution x_0 of S choose a real number $y_0 = f(x_0)$ such that (x_0,y_0) is a real solution of T. Then

Every real solution of S is a solution of $f(x) = f(x)$. (3)

Every real solution of $S \cup \{y = f(x)\}$ is a solution of T. (4)

By Theorem 4, (3) and (4) also hold for hyperreal numbers. Let x_1 be a hyperreal solution of S. By (3), $f(x_1)$ is defined; let $y_1 = f(x_1)$. By (4), (x_1,y_1) is a hyperreal solution of T. Thus x_1 is a partial hyperreal solution of T. This shows that (i) implies (iii).

(iii) trivially implies (ii). Assume (ii), and let x_0 be a real solution of S. Suppose x_0 is not a partial real solution of T. Let $T(x_0,y)$ be the system of formulas obtained by replacing the variable x by the constant x_0 in T. Then $T(x_0,y)$ has no real solutions. By Theorem 1, $T(x_0,y)$ has no hyperreal solutions. But then x_0 is not a partial hyperreal solution of T, contradicting our assumption (ii). We conclude that x_0 is a partial real solution of T, so (ii) implies (i). ⊣

Axioms A-D above are sufficient for the development of calculus. For pedagogical reasons, a different but equivalent set of axioms was given in Elementary Calculus. The Completeness Property in Axiom A is difficult for beginning calculus students to understand or use. It is always possible and usually easier to use the Standard Part Theorem. It turns out that given Axioms B-D, Axiom A (R is a complete ordered field) is equivalent to the assumption that R is an Archimedean ordered field and the Standard Part Theorem holds. We shall now state the axioms from Elementary Calculus (with the original numbering) and prove their equivalence with Axioms A-D of this monograph.

AXIOMS I-III

R is an Archimedean ordered field.

AXIOMS I*-III*

R* is a proper ordered field extension of R. (Same as Axiom B.)

AXIOM IV* (Standard Part Axiom)

Every finite $x \in R^*$ is infinitely close to a unique $r \in R$.

AXIOM V* (Function Axiom)

(Same as Axiom C.)

AXIOM VI* (Solution Axiom)

(Same as Axiom D.)

THEOREM 6

Axioms A-D are equivalent to Axioms I-III and I*-VI*.

Proof We already have shown that Axioms A-D imply Axioms I-III and I*-VI*. Assume that $(R, R^*, *)$ satisfies Axioms I-III and I*-VI*. We must prove that R is complete. Let X be a nonempty subset of R with an upper bound. Let

$$f(y) = \begin{cases} 1 & \text{if } y \text{ is an upper bound of } X \\ 0 & \text{otherwise.} \end{cases}$$

We shall find a real number b such that $f(x) = 0$ for all real $x < b$, and $f(y) = 1$ for all real $y > b$. Since X is nonempty and has an upper bound, there are real numbers a,c with

$$a < c, \ f(a) = 0, \ f(c) = 1.$$

Let t be any positive real number and consider the points

$$a, \ a + t, \ a + 2t, \ \ldots, \ a + nt, \ \ldots .$$

By the Archimedean Axiom there is an n such that

$$(c - a)/t < n, \ c < a + nt .$$

Therefore there is a least positive integer n_0 such that $f(a + n_0 t) = 1$. Hence

$$f(a + n_0 t - t) = 0, \ f(a + n_0 t) = 1 .$$

It follows that every real solution of

$$0 < t \tag{5}$$

is a partial real solution of

$$f(u - t) = 0, \ f(u) = 1 . \tag{6}$$

Let t_1 be positive infinitesimal. By the Partial Solution Theorem t_1 is a partial hyperreal solution of (6). So there is a hyperreal number u_1 with

$$f(u_1 - t_1) = 0, \ f(u_1) = 1 .$$

Let $b = st(u_1)$. We show that for any real $x < b$ and $y > b$, $f(x) = 0$ and $f(y) = 1$. Since every real solution of

$$x < z, \ f(z) = 0$$

is a solution of $f(x) = 0$, we see that $f(x_1) = 0$ for all hyperreal $x_1 < u_1 - t_1$. In particular, if x is real and $x < b = st(u_1)$, then $x < u_1 - t_1$ so $f(x) = 0$. A similar argument shows that $f(y) = 1$ for all real $y > b$. It follows that b is the least upper bound of X. ⊣

1D NATURAL EXTENSIONS OF SETS (§3.3)

Our axioms provide a natural extension f^* of each real function f. We now define the natural extension of a set of reals. Later we shall extend our discussion to relations on the reals.

DEFINITION

Let Y be a set of reals, $Y \subseteq R$. The natural extension of Y is the subset $Y^* \subseteq R^*$ such that every system of formulas which has Y as its set of real solutions has Y^* as its set of hyperreal solutions.

We first show that every set $Y \subseteq R$ has a unique natural extension. By the Solution Axiom, any two systems of formulas having the same set Y of real solutions have the same set Y^* of hyperreal solutions. Thus to show Y^* exists uniquely we need only give an example of a system of formulas whose set of real solutions is Y. The simplest such example is the equation

$$C_Y(x) = 1$$

where C_Y is the characteristic function of Y defined by

$$C_Y(x) = \begin{cases} 1 & \text{if } x \in Y, \\ 0 & \text{if } x \notin Y. \end{cases}$$

Given a term τ and a set $Y \subseteq R$, we shall sometimes write $\tau \in Y$ for the formula $C_Y(\tau) = 1$, and $\tau \notin Y$ for the formula $C_Y(\tau) = 0$.

EXAMPLES Since $[a,b] = \{x \in R: a \leq x, x \leq b\}$,

$$[a,b]^* = \{x \in R^*: a \leq x, x \leq b\}.$$

Since $(a,b) = \{x \in R: a < x, x < b\}$,

$$(a,b)^* = \{x \in R^*: a < x, x < b\}.$$

THEOREM 1

Let $X, Y \subseteq R$.

(i) $X^* \cap R = X$.

(ii) The natural extension mapping preserves Boolean operations, i.e.,

$$(X \cap Y)^* = X^* \cap Y^*,$$
$$(X \cup Y)^* = X^* \cup Y^*,$$
$$(X - Y)^* = X^* - Y^*,$$
$$X \subseteq Y \text{ if and only if } X^* \subseteq Y^*.$$

(iii) For any real function f of one variable,

$$(\text{domain } f)^* = \text{domain}(f^*),$$
$$(\text{range } f)^* = \text{range}(f^*).$$

Proof of (iii) Let X be the domain of f. Then X is the set of all real solutions of the equation

$$f(x) = f(x). \qquad (1)$$

Hence X^* is the set of all hyperreal solutions of (1), and X^* is the domain of f^*.

Let Y be the range of f. For each $y \in Y$ choose a real number $x = g(y)$ such that $f(x) = y$. Then Y is the set of all real solutions of the equation

$$f(g(y)) = y, \qquad (2)$$

so Y^* is the set of all hyperreal solutions of (2). It follows that

$$Y^* \subseteq \text{range}(f^*).$$

Moreover, every real and hence hyperreal solution of the equation $f(x) = y$ is a solution of (2). Therefore

$$\text{range}(f^*) \subseteq Y^*.$$

We shall now explain the connection between hyperreal numbers and topological properties of sets of reals. By a real neighborhood of a point x we mean an open interval of the form

$$(x - r, x + r)$$

where r is a positive real number. x belongs to the interior of Y if some neighborhood of x is included in Y. An open set is a set which is equal to its interior. x belongs to the closure of Y if every neighborhood of x meets Y. A closed set is a set Y which is equal to its closure.

THEOREM 2

Let $c \in R$, $Y \subseteq R$. Y includes a neighborhood of c (i.e., c is in the interior of Y) if and only if Y* includes the monad of c.

Proof Suppose Y contains a neighborhood $(c - r, c + r)$ of c. Then every real solution of

$$|c - x| < r$$

belongs to Y. Hence every hyperreal solution belongs to Y*, and thus the monad of c is included in Y.

Now suppose Y does not contain a neighborhood of c. Then each real solution of

$$x > 0 \tag{3}$$

is a partial real solution of

$$y \notin Y, \ |c - y| < x. \tag{4}$$

Let x_1 be positive infinitesimal. Since x_1 is a hyperreal solution of (3) it is a partial hyperreal solution of (4), so there is a hyperreal y_1 with

$$y_1 \notin Y^*, \ |c - y_1| < x_1 .$$

Then y_1 belongs to the monad of c but not to Y*. ⊣

COROLLARY 3 The closure of a set $Y \subseteq R$ of reals is equal to the set

$$\{st(x): x \text{ is finite and } x \in Y^*\}.$$

Thus Y is closed if and only if for all finite $x \in Y^*$, $st(x) \in Y$.

<u>Proof</u> Let $Z = R - Y$, so $Z^* = R^* - Y^*$. The following are equivalent:

$c \in$ closure of Y.
$c \notin$ interior of Z.
Monad(c) is not included in Z^*.
There is an $x \in Y^*$ such that $x \approx c$.
$c = st(x)$ for some finite $x \in Y^*$. ⊣

COROLLARY 4 A real function f is defined at every point of some neighborhood of c if and only if f^* is defined at every point of the monad of c.

<u>Proof</u> By Theorems 1(iii) and 2. ⊣

A set $Y \subseteq R$ is <u>bounded</u> if Y is included in some closed interval [a,b].

THEOREM 5

Let $Y \subseteq R$. Then Y is bounded if and only if every element of Y^* is finite.

<u>Proof</u> If Y is bounded, $Y \subseteq [a,b]$, then every element $y \in Y$ is a solution of

$$a \le y,\ y \le b, \tag{5}$$

therefore every $y_1 \in Y^*$ is a solution of (5) and hence is finite.

Now suppose Y is not bounded. Then either Y has no upper bound or no lower bound, say Y has no upper bound. Thus each real number x is a partial real solution of

$$x < y,\ y \in Y. \tag{6}$$

Let x_1 be a positive infinite hyperreal number. x_1 is a partial solution of (6), so there is a y_1 such that $x_1 < y_1$, $y_1 \in Y^*$. Thus y_1 is a positive infinite element of Y^*. ⊣

We now extend our discussion to relations on the reals. By a real relation in n variables we mean a subset of R^n.

DEFINITION

Let Y be a real relation in n variables. The natural extension of Y is the subset $Y^* \subseteq R^{*n}$ such that every system of formulas which has Y as its set of real solutions has Y* as its set of hyperreal solutions.

The distance between two points

$$x = (x_1,\dots,x_n),\ y = (y_1,\dots,y_n)$$

of R^n or R^{*n} is defined by

$$|x - y| = \sqrt{(x_1 - y_1)^2 + \dots + (x_n^2 - y_n^2)} .$$

A real neighborhood of a real point $x \in R^n$ is a set of the form

$$N_r(x) = \{y \in R^n : |x - y| < r\},\ 0 < r \in R.$$

We say that y is infinitely close to x, $y \approx x$, if $|x - y|$ is infinitesimal. The monad of x is defined as the set

$$\text{monad}(x) = \{y \in R^{*n} : x \approx y\}.$$

x is finite if each x_i is finite.

With the above definitions, all the results in this section hold for subsets of R^n as well as for subsets of R.

1E A SIMPLE CONSTRUCTION OF THE HYPERREAL NUMBERS

We are now ready to construct the hyperreal numbers. We shall prove that there exists a triple (R,R*,*) which satisfies Axioms A-D. When this is done, it will follow that any statement about the real numbers which follows from the axioms is true of the real numbers. The hyperreal numbers can be regarded as a tool which is constructed in order to facilitate the study of the real numbers.

The construction we describe is a special case of the proof of the Compactness Theorem in mathematical logic. It closely resembles the proof that a field has an extension which contains a root of a given polynomial. In Chapter 1* we shall give a second, more sophisticated construction of the hyperreal numbers, called the ultrapower construction.

Our construction will use a form of the Axiom of Choice called Zorn's Lemma.

A nonempty set of sets X is called a <u>chain</u> if for any two sets $x,y \in X$, either $x \subseteq y$ or $y \subseteq x$.

ZORN'S LEMMA Let Y be a nonempty set of sets such that for any chain $X \subseteq Y$, the union of X belongs to Y. Then Y has a maximal element y, that is, a set $y \in Y$ such that no member of Y properly contains y.

EXISTENCE THEOREM

Let R be the ordered field of real numbers. There is an ordered field extension R* of R and a mapping * from real functions to hyperreal functions (i.e., functions on R*) such that Axioms A-D hold.

<u>Proof</u> Our plan is to find an infinite set M of formulas $\phi(x)$ which describes all properties of a positive infinitesimal x, and to build R* out of this set of formulas. A set of formulas (equations and inequalities) with the single variable x is said to be <u>consistent</u> if every finite subset has a real solution. The union of a chain of consistent sets of formulas is always consistent. Therefore by Zorn's Lemma there is a set of formulas M such that:

$$\{0 < x,\ x < \frac{1}{2},\ x < \frac{1}{3},\ x < \frac{1}{4},\ \ldots\} \subseteq M\ ; \tag{1}$$

$$\text{M is maximal consistent.} \tag{2}$$

Let T(M) be the set of all terms $\tau(x)$ such that $\tau(x) = \tau(x) \in M$. Intuitively, T(M) is the set of terms which are defined modulo M. Two terms $\sigma(x)$ and $\tau(x)$ in T(M) are said to be <u>congruent modulo</u> M if $\sigma(x) = \tau(x) \in M$. Using (2) one can check that congruence is an equivalence relation on T(M). Let ε be the set of all terms congruent to the term x. Given $\tau(x) \in T(M)$, let $\tau[\varepsilon]$ be the set of all terms congruent to $\tau(x)$, that is, $\tau[\varepsilon]$ is the congruence class of $\tau(x)$. Then

$$\sigma[\varepsilon] = \tau[\varepsilon] \quad \text{iff} \quad \sigma(x) = \tau(x) \in M .$$

The hyperreal structure $(R, R^*, *)$ is defined as follows. R^* is the set

$$R^* = \{\tau[\varepsilon] : \tau(x) \in T(M)\} .$$

Thus a hyperreal number $\tau[\varepsilon]$ is a congruence class of terms modulo M. We identify each real number $r \in R$ with the congruence class of r,

$$\sigma[\varepsilon] = r \quad \text{iff} \quad \sigma(x) = r \in M .$$

Thus $R \subseteq R^*$. The order relation $\leq^*$ on R^* is defined by

$$\sigma[\varepsilon] \leq^* \tau[\varepsilon] \quad \text{iff} \quad \sigma(x) \leq \tau(x) \in M .$$

The natural extension of a real function f of one variable is defined by

$$f^*(\tau[\varepsilon]) = f(\tau(x))[\varepsilon] \text{ if } f(\tau(x)) \in T(M) ,$$

$$f^*(\tau[\varepsilon]) \text{ is undefined if } f(\tau(x)) \notin T(M) .$$

Using the maximal consistency (2) of M, it can be shown that $f^*(\tau[\varepsilon])$ depends only on the congruence class $\tau[\varepsilon]$ and not on the term $\tau(x)$. The natural extension of a real function of two or more variables is defined in a similar way.

This completes the construction of $(R, R^*, *)$. Axiom A holds because R is the ordered field of real numbers. The Function Axiom C holds by our construction of f^*. We shall verify Axioms B and D.

Axiom B states that R^* is a proper ordered field extension of R. By condition (1), ε is a positive infinitesimal, so R^* is a proper extension of R. As an illustration of the field axioms we prove the Commutative Law for R^*. Let $\sigma[\varepsilon]$, $\tau[\varepsilon] \in R^*$. Then M remains consistent when the equation

$$\sigma(x) + \tau(x) = \tau(x) + \sigma(x)$$

is added to M. By maximal consistency this equation belongs to M. Thus $\sigma + \tau$ is congruent to $\tau + \sigma$, and

$$\sigma[\varepsilon] +^* \tau[\varepsilon] = (\sigma + \tau)[\varepsilon] = (\tau + \sigma)[\varepsilon]$$
$$= \tau[\varepsilon] +^* \sigma[\varepsilon].$$

To prove that Axiom D holds, we must show that two finite systems of formulas with the same real solutions have the same hyperreal solutions. To keep the notation simple we consider formulas with one variable u. By repeated use of the definition of f*, one can show that for every term $\sigma(u)$ and hyperreal number $\tau[\varepsilon]$ for which $\sigma(\tau[\varepsilon])$ is defined,

$$\sigma(\tau[\varepsilon]) = \sigma(\tau(x))[\varepsilon] \ . \tag{3}$$

Also, $\sigma(\tau[\varepsilon])$ is defined iff $\sigma(\tau(x))[\varepsilon]$ is defined. It follows that for every formula $\phi(u)$ and hyperreal number $\tau[\varepsilon]$,

$$\phi(\tau[\varepsilon]) \text{ is true iff } \phi(\tau(x)) \in M \ . \tag{4}$$

Suppose the two systems of formulas

$$\Phi(u) = \{\phi_1(u), \ldots, \phi_m(u)\} \ ,$$

$$\Psi(u) = \{\psi_1(u), \ldots, \psi_n(u)\}$$

have the same real solutions. Let $\tau[\varepsilon]$ be a hyperreal solution of $\Phi(u)$. Then by (4),

$$\Phi(\tau(x)) = \{\phi_1(\tau(x)), \ldots, \phi_m(\tau(x))\} \subseteq M.$$

Since $\Phi(u)$ and $\Psi(u)$ have the same real solutions u, $\Phi(\tau(x))$ and $\Psi(\tau(x))$ have the same real solutions x. Since M is consistent, for every finite subset $M_0 \subseteq M$ the system $M_0 \cup \Phi(\tau(x))$ has a real solution. Therefore $M_0 \cup \Psi(\tau(x))$ has a real solution, hence $M \cup \Psi(\tau(x))$ is consistent. By maximal consistency,

$$\Psi(\tau(x)) \subseteq M,$$

and using (4) again we conclude that $\tau[\varepsilon]$ is a hyperreal solution of $\Psi(u)$. Our proof is complete. ⊣

1F APPENDIX. ALGEBRA OF THE REAL NUMBERS (§1.1)

This appendix is a summary of some basic notions from algebra which lead up to the characterization of the real number system (Theorem 3). The details are developed in many undergraduate texts on modern algebra.

RING A (commutative) ring is a structure $(R, 0, +, -, \cdot)$ such that the following axioms hold for all $a,b,c \in R$.

Closure Laws	$0 \in R$, $a + b \in R$, $-a \in R$, $a \cdot b \in R$
Commutative Laws	$a + b = b + a$, $a \cdot b = b \cdot a$

Associative Laws	$a + (b + c) = (a + b) + c$, $a\cdot(b\cdot c) = (a\cdot b)\cdot c$
Identity Law	$a + 0 = a$
Inverse Law	$a + (-a) = 0$
Distributive Law	$a\cdot(b + c) = a\cdot b + a\cdot c$

SUBRING A subring of a ring R is a subset S of R which satisfies the Closure Laws. It follows that S is itself a ring.

IDEAL An ideal in a ring R is a subring I of R such that whenever $a \in I$ and $r \in R$, $a\cdot r \in I$.

COSET Given a ring R and a subring S, the coset of an element $r \in R$ modulo S is the set

$$\text{coset}(r) = \{r + s\colon s \in S\}.$$

EQUIVALENCE RELATION An equivalence relation on a set A is a binary relation $\equiv$ such that for all a,b,c A,

$$a \equiv a,$$
$$a \equiv b \text{ implies } b \equiv a,$$
$$a \equiv b \text{ and } b \equiv c \text{ implies } a \equiv c.$$

THEOREM 1

If S is a subring of R then the relation $a - b \in S$ is an equivalence relation on R. Moreover,

$$\text{If } a - b \in S \text{ then coset}(a) = \text{coset}(b),$$
$$\text{If } a - b \notin S \text{ then coset}(a) \cap \text{coset}(b) = \emptyset.$$

HOMOMORPHISM A homomorphism from a ring R into a ring S is a function $h\colon R \to S$ such that for all $a,b \in R$,

$$h(a + b) = h(a) + h(b),\ h(-a) = -h(a),\ h(a\cdot b) = h(a)\cdot h(b)\ .$$

ISOMORPHISM An isomorphism from a ring R to a ring S is a homomorphism h: R → S such that h maps R one to one onto S.

FIELD A field is a structure $(F, 0, 1, +, -, \cdot, ^{-1})$ such that $(F, 0, +, -, \cdot)$ is a commutative ring and the following hold for all $a \neq 0$ in F.

Closure Laws	$1 \in F$, $a^{-1} \in F$
Nontriviality	$1 \neq 0$
Identity Law	$1 \cdot a = a$
Inverse Law	$a \cdot (a^{-1}) = 1$

SUBFIELD A subfield of a field F is a subset G of F which satisfies the Closure Laws. It follows that G itself is a field.

PROPER FIELD EXTENSION A proper field extension of G is a field F such that G is a subfield of F and $G \neq F$.

ORDERED FIELD An ordered field is a field $(F, 0, 1, +, -, \cdot, ^{-1})$ with a binary relation < such that the following axioms hold for all $a,b,c \in F$.

Transitive Law	If $a < b$ and $b < c$ then $a < c$
Trichotomy Law	Exactly one of the relations $a < b$, $a = b$, $b < a$ hold
Sum Law	If $a < b$ then $a + c < b + c$
Product Law	If $a < b$ and $0 < c$ then $a \cdot c < b \cdot c$

EXAMPLES

Z, the ring of integers
Q, the ordered field of rational numbers
R, the ordered field of real numbers
C, the field of complex numbers
GF(2), the field with two elements 0,1, where $1 + 1 = 0$

Z is a subring of Q, Q is a subfield of R, and R is a subfield of C. The set of even integers is an ideal in Z. The mapping h: Z → GF(2) given by

$$h(n) = \begin{cases} 0 & \text{if } n \text{ is even} \\ 1 & \text{if } n \text{ is odd} \end{cases}$$

is a homomorphism of Z onto the two element field.

THEOREM 2

The following algebraic rules hold in every ordered field.

$$a(-b) = (-a)b = -(ab)$$

$$a \cdot 0 = 0, \quad 0 < 1$$

$$-(-a) = a, \quad (a^{-1})^{-1} = a, \quad (a \neq 0)$$

$$-(a - b) = b - a, \quad (a/b)^{-1} = b/a, \quad (a,b \neq 0)$$

$$\frac{a}{b} + \frac{c}{d} = \frac{ad + bc}{bd}, \quad \frac{a}{b} \cdot \frac{c}{d} = \frac{ac}{bd} \qquad (b,d \neq 0)$$

If $a < b$ then $-b < -a$

If $0 < a < b$ then $b^{-1} < a^{-1}$

$$|-a| = |a|, \quad |ab| = |a|\,|b|, \quad \frac{|a|}{|b|} = \left|\frac{a}{b}\right|, \quad (b \neq 0),$$

$$\text{where } |a| = \begin{cases} a & \text{if } 0 \leq a \\ -a & \text{if } a < 0 \end{cases}$$

As usual, $a \leq b$ means $a < b$ or $a = b$. n is the field element

$$n = \underbrace{1 + 1 + \ldots + 1}_{n \text{ times}}.$$

ARCHIMEDEAN ORDERED FIELD An ordered field F is Archimedean if each element $a \in F$ is less than some positive integer,

$$a < \underbrace{1 + \ldots + 1}_{n} \text{ for some } n.$$

COMPLETE ORDERED FIELD An ordered field F is complete if every nonempty subset $X \subseteq F$ which has an upper bound in F has a least upper bound in F.

ORDER ISOMORPHISM An order isomorphism is an isomorphism h from an ordered field F onto an ordered field G such that for any $a,b \in F$, $a < b$ if and only if $h(a) < h(b)$.

THEOREM 3

There is up to order isomorphism exactly one complete ordered field, called the field of <u>real numbers</u>.

THEOREM 4

Every complete ordered field is Archimedean. In fact, an ordered field G is Archimedean if and only if it is order isomorphic to a subfield of the field of real numbers.

Since we often refer to intervals in the real line, we shall review the definition and notation here.

INTERVAL A real interval is a set I of real numbers such that if $a,b \in I$ and $a < c < b$, then $c \in I$.

It follows from the completeness property that every interval is of one of the types below:

Bounded closed intervals

$$[a,b] = \{x: a \le x \le b\}.$$

Bounded open intervals

$$(a,b) = \{x: a < x < b\}.$$

Bounded half-open intervals

$$[a,b), \ (a,b].$$

Unbounded open intervals

$$(a,\infty) = \{x: a < x\},$$
$$(-\infty,b) = \{x: x < b\},$$
$$(-\infty,\infty) = \text{real line}.$$

Unbounded half-open intervals

$$[a,\infty), \ (-\infty,b].$$

LOGIC AND SUPERSTRUCTURES (OPTIONAL) 1*

This chapter provides a link between the simple treatment of infinitesimal calculus in the text *Elementary Calculus* and the more advanced treatment of infinitesimal analysis found in the literature. The material in this chapter is not used elsewhere in the book, and is not needed as background for teaching calculus from *Elementary Calculus*. It is aimed at mathematicians who wish to go more deeply into the subject.

1A* THE ELEMENTARY EXTENSION PRINCIPLE

We shall show here that the Solution Axiom is equivalent to an apparently stronger statement called the Elementary Extension Principle. It says that the real numbers (with all real functions and relations) satisfy the same sentences of first order logic as the hyperreal numbers (with the natural extensions of all real functions and relations). Before stating the result precisely we start from the beginning and define the notion of a sentence of first order logic.

In first order logic we start with a set of symbols, called a *language*, appropriate for the structure under consideration. In this case we introduce a language L for the real number system with the following uncountable collection of symbols:

A symbol $\overline{c}$ for each real constant c,
A symbol $\overline{f}$ for each real function f of n variables,
A symbol $\overline{P}$ for each real relation P of n variables.

In addition, L has the following logical symbols common to all first order languages:

Variables $v_1, v_2, v_3, \ldots$

Connectives $\neg$ (not), $\wedge$ (and), $\vee$ (or), $\rightarrow$ (implies), $\leftrightarrow$ (if and only if)

Quantifiers $\forall$ (for all), $\exists$ (there exists)

Parentheses and commas.

For simplicity we identify each real constant c, function f, and relation P, with its symbol $\overline{c}$, $\overline{f}$, and $\overline{P}$. In §1B we defined the notion of a term. We repeat the definition here. A term is a finite sequence of symbols built according to the following rules:

(T1) Every variable is a term.

(T2) Every constant is a term.

(T3) If $\tau_1,\ldots,\tau_n$ are terms and f is a real function of n variables, then $f(\tau_1,\ldots,\tau_n)$ is a term.

When we replace each variable v_i in a term $\tau(v_1,\ldots,v_n)$ by a constant c_i we obtain a constant term, which is either equal to some real number or is undefined.

In §1B we called an equation or inequality between two terms a formula,

$$\tau = \sigma,\ \tau \neq \sigma,\ \tau \leq \sigma,\ \tau < \sigma\ \ \tau \geq \sigma,\ \tau > \sigma\ .$$

The equations and inequalities are special cases of a broader class which we call the atomic formulas of L. If P is a real relation of n variables and $\tau_1,\ldots,\tau_n$ are terms, then

$$P(\tau_1,\ldots,\tau_n)$$

is called an atomic formula of L.

The set of all (first order) formulas of L is defined as follows.

(F1) Every atomic formula of L is a formula of L.

(F2) If ϕ, ψ are formulas of L, so are

$$(\neg\phi),\ (\phi \wedge \psi),\ (\phi \vee \psi),\ (\phi \rightarrow \psi),\ (\phi \leftrightarrow \psi).$$

(F3) If ϕ is a formula of L and v_n is a variable, then

$$(\forall v_n \phi), \quad (\exists v_n \phi)$$

are formulas of L.

For example, omitting unnecessary parentheses, the following formula of L is the standard ε,δ condition for f to be continuous at x. For readability we use ε, δ, x, y for variables instead of v_1, v_2, v_3, v_4.

$$\forall \varepsilon(\varepsilon > 0 \rightarrow \exists \delta(\delta > 0 \wedge \forall y(|x - y| < \delta \rightarrow |f(x) - f(y)| < \varepsilon))) \ .$$

In this example, ε, δ, and y are <u>bound variables</u>, while x is a <u>free variable</u> because it is not in a quantifier. A first order formula with no free variables is called a <u>sentence</u>. Thus whenever all the free variables in a first order formula are replaced by constants, the result is a sentence.

The statement that two systems of equations and inequalities $S = \{\sigma_1,\ldots,\sigma_m\}$ and $T = \{\tau_1,\ldots,\tau_n\}$ have the same real solutions is expressed by a sentence of L of the form

$$\forall v_1 \ldots \forall v_\ell (\sigma_1 \wedge \ldots \wedge \sigma_m \leftrightarrow \tau_1 \wedge \ldots \wedge \tau_n). \qquad \text{(A)}$$

The statement that every real solution of S is a partial real solution of T is expressed by a sentence of L of the form

$$\forall v_1 \ldots \forall v_k (\sigma_1 \wedge \ldots \wedge \sigma_m \rightarrow \exists v_{k+1} \ldots \exists v_\ell \tau_1 \wedge \ldots \wedge \tau_n). \qquad \text{(B)}$$

Each sentence ϕ of L is either true or false in the real number system. The notion of a true sentence is a precise mathematical concept which is defined by induction on the complexity of a sentence, and corresponds to the intuitive notion.

DEFINITION

(i) An atomic sentence $P(\tau_1,\ldots,\tau_n)$ is true if each of the constant terms $\tau_1,\ldots,\tau_n$ is defined and the n-tuple of values belongs to the relation P.

(ii) If ϕ and ψ are sentences, then the truth values of the combinations of ϕ and ψ by connectives are obtained from the truth values of ϕ and ψ using the following table.

ϕ	ψ	$\neg\phi$	$\phi\wedge\psi$	$\phi\vee\psi$	$\phi\rightarrow\psi$	$\phi\leftrightarrow\psi$
T	T	F	T	T	T	T
T	F	F	F	T	F	F
F	T	T	F	T	T	F
F	F	T	F	F	T	T

(iii) The sentence $\forall x\phi(x)$ is true if and only if $\phi(c)$ is true for all constants $c \in R$.

(iv) The sentence $\exists x\phi(x)$ is true if and only if $\phi(c)$ is true for some constant $c \in R$.

The notation $R \models \phi$ means ϕ is true in R.

Note that we treat = and $\neq$ as binary relations, and that the sentence $\tau \neq \sigma$ is different than $\neg\tau = \sigma$. In fact, $R \models \tau \neq \sigma$ if and only if τ and σ are both defined but have different values, while $R \models \neg\tau = \sigma$ if and only if it is not the case that τ and σ are both defined and equal.

We now introduce a second language L* for the hyperreal number system which has a symbol for each hyperreal constant, function, and relation. The truth value of a sentence of L* in R* is defined as before but with the quantifiers ranging over R* instead of R.

DEFINITION

Given a formula ϕ of L, the *-transform of ϕ is the formula ϕ^* of L* obtained by replacing each real function f and relation P occurring in ϕ by its natural extension f* and P*.

Thus the *-transform of the formula expressing continuity of f at x is the formula

$$\forall\varepsilon(\varepsilon >^* 0 \rightarrow \exists\delta(\delta >^* 0 \wedge \forall y(|x -^* y|^* <^* \delta \rightarrow |f^*(x) -^* f^*(y)|^* <^* \varepsilon))).$$

The relation $x \approx y$ is expressed by the following formula of L*:

$$\forall z(R(z) \wedge 0 <^* z \rightarrow |x -^* y|^* <^* z) .$$

This formula is not the *-transform of any formula of L because it involves the set R, which is a subset of R* but is not the natural extension of any set of reals.

Similarly the property "x is finite" is expressed by

$$\exists z(R(z) \wedge |x|^* <^* z),$$

which again is not the *-transform of any formula of L.

The *-transform of a <u>sentence</u> of L is always a sentence of L*. We can now state the Elementary Extension Principle and show that it follows from our Axioms A-D for hyperreal numbers.

ELEMENTARY EXTENSION PRINCIPLE

For every sentence ϕ of L, ϕ is true in R if and only if its *-transform ϕ^* is true in R*.

Notice that the Solution Axiom is just the special case of the Elementary Extension Principle where ϕ is a sentence of the form (A) above. The Partial Solution Theorem in §1C is the special case where ϕ is a sentence of the form (B) above.

<u>Proof of the Elementary Extension Principle</u> For each formula $\phi(v_1,\ldots,v_n)$ of L with at most the free variables $v_1,\ldots,v_n$, let C_ϕ be the corresponding characteristic function,

$$C_\phi(a_1,\ldots,a_n) = \begin{cases} 1 & \text{if } \phi(a_1,\ldots,a_n) \text{ holds in R} \\ 0 & \text{otherwise.} \end{cases}$$

Thus

$$R \models \forall v_1 \ldots \forall v_n(\phi(v_1,\ldots,v_n) \leftrightarrow C_\phi(v_1,\ldots,v_n) = 1).$$

We shall prove by induction on the complexity of formulas that for each formula $\phi(v_1,\ldots,v_n)$ of L,

$$R^* \models \forall v_1 \ldots \forall v_n(\phi^*(v_1,\ldots,v_n) \leftrightarrow C^*_\phi(v_1,\ldots,v_n) = 1). \quad (1)$$

C^* means $(C_\phi)^*$, not $C_{(\phi^*)}$. Replacing variables by real constants, it will then follow that the statements below are equivalent:

$$R \models \phi(a_1,\ldots,a_n)$$

$$R \models C_\phi(a_1,\ldots,a_n) = 1$$

$$R^* \models C^*_\phi(a_1,\ldots,a_n) = 1 \quad \text{(By Theorem 2 of §1C)}$$

$$R^* \models \phi^*(a_1,\ldots,a_n).$$

First let $\phi(v_1,\ldots,v_n)$ be an atomic formula $P(\tau_1,\ldots,\tau_k)$. Let C_P be the characteristic function of $P(y_1,\ldots,y_k)$. By the definition of P^* in §1D,

$$R^* \models \forall y_1\ldots\forall y_k(P^*(y_1,\ldots,y_k) \leftrightarrow C^*_P(y_1,\ldots,y_k) = 1),$$

and since $\phi(v_1,\ldots,v_n)$ is $P(\tau_1,\ldots,\tau_k)$,

$$R^* \models \forall v_1\ldots\forall v_n(\phi^*(v_1,\ldots,v_n) \leftrightarrow C^*_P(\tau^*_1,\ldots,\tau^*_k) = 1).$$

Moreover, in R we have

$$R \models \forall v_1\ldots\forall v_n(C_P(\tau_1,\ldots,\tau_k) = 1 \leftrightarrow C_\phi(v_1,\ldots,v_n) = 1).$$

Then by the Solution Axiom,

$$R^* \models \forall v_1\ldots\forall v_n(C^*_P(\tau^*_1,\ldots,\tau^*_k) = 1 \leftrightarrow C^*_\phi(v_1,\ldots,v_n) = 1).$$

Combining the above formulas, we see that (1) holds for ϕ.

To shorten our induction we use the fact that every formula of L is equivalent to a formula which is built up using only the connectives $\wedge$ and $\neg$ and the quantifier $\exists$. The other connectives and quantifiers may be treated as abbreviations for longer expressions. In the following we assume that (1) holds for $\phi(v_1,\ldots,v_n)$ and $\psi(v_1,\ldots,v_n)$, and prove that (1) also holds for $\neg\phi$, $\phi\wedge\psi$, and $\exists v_n\phi$.

We first consider $\neg\phi$. In R we have

$$R \models \forall v_1\ldots\forall v_n(C_\phi(v_1,\ldots,v_n) = C_\phi(v_1,\ldots,v_n)),$$

$$R \models \forall v_1\ldots\forall v_n(C_\phi(v_1,\ldots,v_n) \neq 1 \leftrightarrow C_{\neg\phi}(v_1,\ldots,v_n) = 1).$$

By the Solution Axiom the *-transforms hold in R*. Therefore

$$R^* \models \forall v_1\ldots\forall v_n(\neg\, C^*_\phi(v_1,\ldots,v_n) = 1$$
$$\leftrightarrow C^*_{\neg\phi}(v_1,\ldots,v_n) = 1).$$

Using the hypothesis (1) for ϕ, we have

$$R^* \models \forall v_1 \ldots \forall v_n (\neg \phi^*(v_1,\ldots,v_n) \leftrightarrow \neg C^*_\phi(v_1,\ldots,v_n) = 1).$$

It follows that

$$R^* \models \forall v_1 \ldots \forall v_n (\neg \phi^*(v_1,\ldots,v_n) \leftrightarrow C^*_{\neg\phi}(v_1,\ldots,v_n) = 1)$$

so (1) holds for $\neg\phi$.

To verify (1) for $\phi \wedge \psi$ we observe that

$$R \models \forall v_1 \ldots \forall v_n (C_{\phi\wedge\psi}(v_1,\ldots,v_n) = C_\phi(v_1,\ldots,v_n) \cdot C_\psi(v_1,\ldots,v_n)),$$

and by the Solution Axiom the *-transform holds in R*.

We now show that $\exists v_n \phi$ satisfies (1). For any real (n - 1)-tuple $a_1,\ldots,a_{n-1}$ such that

$$R \models \exists v_n \phi(a_1,\ldots,a_{n-1},v_n),$$

choose a real number $f(a_1,\ldots,a_{n-1})$ such that

$$R \models \phi(a_1,\ldots,a_{n-1},f(a_1,\ldots,a_{n-1})).$$

Otherwise leave $f(a_1,\ldots,a_{n-1})$ undefined. The real function f of n - 1 variables is called a <u>Skolem function</u> for $\exists v_n \phi$. In R we see that

$$R \models \forall v_1 \ldots \forall v_{n-1} (C_{\exists v_n \phi}(v_1,\ldots,v_{n-1}) = 1 \leftrightarrow C_\phi(v_1,\ldots,v_{n-1},f(v_1,\ldots,v_{n-1})) = 1) \quad (2)$$

and also

$$R \models \forall v_1 \ldots \forall v_n (C_\phi(v_1,\ldots,v_n) = 1 \rightarrow C_{\exists v_n \phi}(v_1,\ldots,v_{n-1}) = 1) . \quad (3)$$

It follows from the Solution Axiom that the *-transforms hold in R*. Thus for any $b_1,\ldots,b_{n-1} \in R^*$, each statement below implies the next.

$$R^* \models \exists v_n \phi^*(b_1,\ldots,b_{n-1},v_n)$$

$$R^* \models \exists v_n (C^*_\phi(b_1,\ldots,b_{n-1},v_n) = 1) \quad \text{by (1) for } \phi,$$

$$R^* \models C^*_{\exists v_n \phi}(b_1,\ldots,b_{n-1}) = 1 \quad \text{by (3)},$$

$$R^* \models C^*_{\phi}(b_1,\ldots,b_{n-1},f^*(b_1,\ldots,b_{n-1})) = 1 \quad \text{by (2)},$$

$$R^* \models \phi^*(b_1,\ldots,b_{n-1},f^*(b_1,\ldots,b_{n-1})) \quad \text{by (1) for } \phi,$$

$$R^* \models \exists v_n \phi^*(b_1,\ldots,b_{n-1},v_n) \ .$$

We conclude that (1) holds for $\exists v_n \phi$. ⊣

In the textbook Elementary Calculus and in the remaining chapters of this monograph we use the Solution Axiom directly instead of using the Elementary Extension Principle which we have just proved. The sentences which arise in beginning calculus are of a very simple form, and it turns out that the Solution Axiom is broad enough to cover the cases of the Elementary Extension Principle which are needed in proofs. In this monograph we shall sometimes go one step beyond the Solution Axiom and use the Partial Solution Theorem, but we have no need for more complicated cases. It is very important pedagogically that we are able to base the course on the familiar concept of a system of equations and inequalities rather than on the general notion of a formula in first order logic. Beginning calculus students do not have the mathematical experience necessary to work with formulas with even three quantifiers, such as the ε,δ definition of continuity. One advantage of introducing the hyperreal numbers at the beginning of the calculus course is that complicated formulas of L can often be replaced by simpler formulas of L*. This is especially true when the hyperreal relations $x \approx y$ and $x = st(y)$ are used. For example, the ε,δ condition for f to be continuous at x,

$$\forall \varepsilon(\varepsilon > 0 \rightarrow \exists \delta(\delta > 0 \wedge \forall y(|x - y| < \delta \rightarrow |f(x) - f(y)| < \varepsilon))),$$

is equivalent to the simpler L* formula

$$\forall y(y \approx x \rightarrow f^*(y) \approx f^*(x)) \ .$$

1B* SUPERSTRUCTURES

In classical analysis one goes beyond the real numbers and real functions. A more appropriate object of study is the superstructure over the real numbers, defined as follows.

DEFINITION

The power set of a set X is the set $\mathcal{P}(X)$ of all subsets of X,

$$\mathcal{P}(X) = \{Y: Y \subseteq X\}.$$

The nth cumulative power set of X is defined recursively by

$$V_0(X) = X,$$

$$V_{n+1}(X) = V_n(X) \cup \mathcal{P}(V_n(X)).$$

Thus

$$X \subseteq V_1(X) \subseteq V_2(X) \subseteq \dots \text{ and } X \in V_1(X) \in V_2(X) \in \dots .$$

The superstructure over X is the union of the cumulative power sets and is denoted by V(X),

$$V(X) = \bigcup_{n=0}^{\infty} V_n(X).$$

It follows that $V_n(X) \in V(X)$ for each n. The superstructure V(X) has a membership relation between elements of $V_n(X)$ and elements of $V_{n+1}(X)$, $n = 0,1,2,\dots$. We treat the elements of X itself as atoms, and assume that $\emptyset \notin X$ and that no $x \in X$ contains any elements of V(X).

Typically one takes the set of atoms X to be the set R of reals, the set C of complex numbers, or some other structure under investigation. We shall concentrate on the case where the set of atoms is R.

The sets Z of integers and N of natural numbers (nonnegative integers) are important subsets of R and elements of V(R). We may define an ordered pair $\langle x,y\rangle$ by

$$\langle x,y\rangle = \{\{x\},\{x,y\}\}$$

and an ordered n-tuple, $n > 2$, as the function

$$\langle x_1,\dots,x_n\rangle = \{\langle 1,x_1\rangle,\dots,\langle n,x_n\rangle\}$$

from $\{1,\dots,n\}$ into R. Thus all ordered n-tuples belong to $V_3(R)$, and all real relations and functions in n variables belong to $V_4(R)$. Function spaces, measures, and all other structures from classical analysis belong to V(R), and even to, say, $V_{100}(R)$.

In order to be in a position to use infinitesimals in more advanced areas of mathematics, we must extend the whole superstructure instead of just the real numbers.

DEFINITION

A superstructure embedding is a one to one mapping * of V(R) into another superstructure V(R*) such that

(i) R is a proper subset of R* and $r^* = r$ for all $r \in R$.

(ii) For $x,y \in V(R)$, $x \in y$ if and only if $x^* \in y^*$.

Notice that by (i), $V(R) \subsetneq V(R^*)$ but * is not the identity embedding on V(R).

To go further we need an analogue of the Elementary Extension Principle for superstructure embeddings. An arbitrary embedding will not do; we want the embedding to preserve all properties of a certain kind. Assume hereafter that $*: V(R) \to V(R^*)$ is a superstructure embedding.

We first form the first order predicate logic with the equality symbol, a binary relation symbol $\in$, and a constant symbol $\bar{a}$ for each element $a \in V(R^*)$. For simplicity we identify a with its constant symbol $\bar{a}$. We call the language $\mathcal{L}$.

DEFINITION

A (bounded) formula of $\mathcal{L}$ is a formula built according to the following rules.

(i) If x and y are variables or constants, $x = y$ and $x \in y$ are formulas (called atomic formulas).

(ii) If ϕ, ψ are formulas, so are $\neg\phi$, $\phi \wedge \psi$, $\phi \vee \psi$, $\phi \to \psi$, $\phi \leftrightarrow \psi$.

(iii) If u, v are variables and ϕ is a formula, then $(\forall u \in v)\phi$ and $(\exists u \in v)\phi$ are formulas.

(iv) If u is a variable, c a constant, and ϕ a formula, then $(\forall u \in c)\phi$ and $(\exists u \in c)\phi$ are formulas.

The adjective bounded refers to the fact that all the quantifiers are bounded. Since we only consider bounded formulas here, we can simply say "formula" for "bounded formula."

A sentence is a formula in which each variable u occurs within the scope of a quantifier of the form $(\forall u \in x)$ or $(\exists u \in x)$ where x is another variable or a constant. Each sentence is either true or false in the superstructure $V(R^*)$. The definition of the relation "ϕ is true in $V(R^*)$" is by induction on the complexity of ϕ. The quantifier clauses are:

$(\exists v \in c)\phi$ is true in $V(R^*)$ iff
$\phi(b)$ is true in $V(R^*)$ for some $b \in c$,

$(\forall v \in c)\phi$ is true in $V(R^*)$ iff
$\phi(b)$ is true in $V(R^*)$ for all $b \in c$.

The notation $V(R^*) \models \phi$ is used in model theory to denote the relation "ϕ is true in $V(R^*)$." Since $V(R^*)$ is the only structure which will be considered in our discussion, we may suppress mention of $V(R^*)$ and simply say "ϕ is true" instead of "ϕ is true in $V(R^*)$."

Remember that $R \subseteq R^*$ and $V(R) \subseteq V(R^*)$. We call the elements of R real numbers, the elements of $V(R) - R$ real sets, and arbitrary elements of $V(R)$ real entities. Similarly, elements of R^*, $V(R^*) - R^*$, and $V(R^*)$ are called hyperreal numbers, sets, and entities respectively. A real formula is a formula all of whose constants are real entities. Since each element of a real set is a real entity, a real sentence has the same meaning in $V(R)$ as it has in $V(R^*)$. The superstructure embedding $*: V(R) \to V(R^*)$ induces a mapping, called the *-transform, from real formulas to hyperreal formulas. The *-transform ϕ^* of a real formula ϕ is defined as the formula obtained by replacing each constant c occurring in ϕ by its image c^*.

For example, the *-transform of the real sentence

$$(\forall x \in R)(x < 0 \vee (\exists y \in R) y \cdot y = x)$$

is the hyperreal sentence

$$(\forall x \in R^*)(x <^* 0^* \vee (\exists y \in R^*) y \cdot^* y = x).$$

DEFINITION

A superstructure embedding $*: V(R) \to V(R^*)$ preserves bounded sentences if for each real sentence ϕ of $\mathcal{L}$, ϕ is true if and only if ϕ^* is true.

LEIBNIZ' PRINCIPLE

$*: V(R) \to V(R^*)$ preserves bounded sentences.

Of course, Leibniz did not formulate his principle in anything like the present form. In fact, first order predicate logic was not available until the work of Frege in the late nineteenth century. The name "Leibniz' Principle" is often used in the literature because Leibniz suggested that the real numbers should be extended to a larger system which has the same elementary properties but contains infinitesimals.

In Section 1D* we shall construct a superstructure embedding which satisfies Leibniz' Principle. We now show that Leibniz' Principle implies the Elementary Extension Principle of the preceding section.

By the elementary part of $V(R)$ we shall mean the set

$$R \cup \bigcup_{n=1}^{\infty} \mathcal{P}(R^n)$$

of all elements of R and all relations (and hence functions) on R. By the elementary part of $*$ we shall mean the restriction of $*$ to the elementary part of $V(R)$. Thus the elementary part of $*$ associates with each real relation P or function f a hyperreal relation P^* or function f^*.

THEOREM 1

If $*: V(R) \to V(R^*)$ is a superstructure embedding which satisfies Leibniz' Principle (preserves bounded sentences), then the elementary part of $*$ satisfies the Elementary Extension Principle.

Proof Given a first order sentence ϕ of L, let ϕ^R be the real sentence of $\mathcal{L}$ obtained by replacing each quantifier $\exists v_n$ or $\forall v_n$ by $(\exists v_n \in R)$ or $(\forall v_n \in R)$. Define ϕ^{R^*} similarly. Then for each sentence ϕ of L,

$$(\phi^*)^{R^*} = (\phi^R)^* .$$

From the definition of truth value we see that the following are equivalent:

$$R \models \phi,\ \phi^R \text{ is true},\ (\phi^R)^* \text{ is true},$$

$$(\phi^*)^{R^*} \text{ is true},\ R^* \models \phi^* . \qquad \dashv$$

COROLLARY 2 Let $*: V(R) \to V(R^*)$ be a superstructure embedding which satisfies Leibniz' Principle. Then the structure $(R, R^*, *_0)$, where $*_0$ is the elementary part of $*$, satisfies Axioms A-D for the hyperreal numbers.

1C* STANDARD, INTERNAL, AND EXTERNAL SETS

In this section we assume that $*: V(R) \to V(R^*)$ is a superstructure embedding which satisfies Leibniz' Principle. We shall adopt the convention of dropping asterisks on terms $f^*(x_1,\ldots,x_n)$ where f is a real function, and on the hyperreal relations $<$, $\leq$.

A number of new distinctions which did not arise in the hyperreal numbers become important in superstructure embeddings. The image of the power set of R will contain some but not all subsets of R^*, that is, $[\mathcal{P}(R)]^*$ will be a proper subset of $\mathcal{P}(R^*)$. For example, the set R of real numbers and the set N of natural numbers are subsets of R^* but do not belong to $[\mathcal{P}(R)]^*$. To see this we note that the $*$-transform of the sentence "every nonempty bounded set in $\mathcal{P}(R)$ has a least upper bound" holds in $V(R^*)$. However, R and N are bounded but have no least upper bound in R^*, so they cannot belong to $[\mathcal{P}(R)]^*$.

It is useful to distinguish between four kinds of entities in $V(R^*)$. An entity $b \in V(R^*)$ is said to be:

<u>real</u>	if $b \in V(R)$,
<u>standard</u>	if $b = a^*$ for some $a \in V(R)$,
<u>internal</u>	if $b \in a^*$ for some $a \in V(R)$,
<u>external</u>	if b is not internal.

REMARKS (i) Every standard entity is internal.

(ii) x is internal if and only if $x \in [V_n(R)]^*$ for some n.

<u>Proof</u> (i) If x is standard, $x = a^*$ for some real a, and since $a \in \mathcal{P}(a)$, $x \in [\mathcal{P}(a)]^*$ so x is internal. ⊣

EXAMPLES <u>Standard and real</u>. Each $r \in R$, each finite subset of R.

<u>External and real</u>. N, R.

<u>Standard but not real</u>. N^*, R^*.

Internal but not standard. Each $c \in R^* - R$, $[a,b]^*$ where $a,b \in R^* - R$, the function $f(x) = \sin(Hx)$ where H is infinite.
External but not real. Monad(0), galaxy(0), the standard part function.

A formula of $\mathcal{L}$ is said to be real, standard, internal, or external if all its constants have the corresponding property. Thus the *-transform of a real formula is a standard formula.

THEOREM 1

A set $X \subseteq R$ is internal if and only if it is finite.

Proof If X is finite, say $X = \{r_1,\ldots,r_n\}$, then

$$(\forall x \in X)(x = r_1 \vee \ldots \vee x = r_n),$$

and hence

$$(\forall x \in X^*)(x = r_1^* \vee \ldots \vee x = r_n^*).$$

Therefore

$$X^* = \{r_1^*,\ldots,r_n^*\} = \{r_1,\ldots,r_n\} = X,$$

so X is internal and in fact standard.

Now suppose X is internal. Then $X \in [\mathcal{P}(R)]^*$. Let $\lhd$ be a well-ordering of R, and let n_1 be a positive infinite hyperinteger. By Leibniz' Principle,

$(\forall Y \in [\mathcal{P}(R)]^*)(\forall n \in N^*)$ (There is a greatest $m \le n$ such that there is a map $f \in [V_3(R)]^*$ of $\{0,\ldots,m\}$ into Y which is increasing in $\lhd^*$).

Therefore there is a greatest $n_0 \le n_1$ with an internal map $f\colon \{0,\ldots,n_0\} \to X$ increasing in $\lhd^*$. Assume X is infinite. Then n_0 must be infinite. On R, $\lhd^*$ is the same as $\lhd$, so f is increasing in $\lhd$. But then the set

$$\{f(n)\colon n \text{ is infinite and } n \le n_0\}$$

is a nonempty set which has no least element, because if n is infinite so is n - 1. This contradiction shows that X is finite. ⊣

A similar argument shows that every real set $Y \in V(R)$ which is internal is both standard and finite. The exact result is that Y is real and internal if and only if $Y \in V(X)$ for some finite $X \subseteq R$.

Several basic notions such as open set, continuity, and differentiability, split into two separate notions when applied to internal functions. Consider an internal function f on R* and a point $c \in R^*$. Let R^+ be the set of positive reals. f is said to be S-continuous at c if it satisfies the real ε,δ condition

$$(\forall \varepsilon \in R^+)(\exists \delta \in R^+)(\forall x \in R^*)(|x - c| < \delta \rightarrow |f(x) - f(c)| < \varepsilon).$$

f is said to be *-continuous at c if it satisfies the hyperreal ε,δ condition,

$$(\forall \varepsilon \in (R^+)^*)(\exists \delta \in (R^+)^*)(\forall x \in R^*)(|x - c| < \delta \rightarrow |f(x) - f(c)| < \varepsilon).$$

It follows from Leibniz' Principle that if f is a standard function and c is real, then f is S-continuous at c if and only if f is *-continuous at c. First we use the principle to show that for each real $\varepsilon,\delta \in R^+$, we may replace $(\forall x \in R^*)$ by $(\forall x \in R)$ in the definition of S-continuous, then we use the principle again to show that S-continuous is equivalent to *-continuous. In general, however, the two notions are different. S-continuous is defined by an external formula and *-continuous by an internal formula. Here are two examples. Let H be positive infinite. The internal function $f(x) = \sin(Hx)$ is everywhere *-continuous and nowhere S-continuous. The internal function

$$g(x) = \begin{cases} 1/H & \text{if } x \in Q^* \\ 0 & \text{if } x \notin Q^*, \end{cases}$$

where Q is the set of rationals, is everywhere S-continuous but nowhere *-continuous.

Some properties of the real numbers cannot be preserved by any superstructure embedding. For example, the field R is Archimedean but the larger field R* is not. The property of being Archimedean cannot be uniformly expressed by a bounded sentence. A field is Archimedean if every element is less than some standard integer. Formally we have

$$(\forall x \in R)(\exists n \in N)x < n, \quad \neg(\forall x \in R^*)(\exists n \in N)x < n.$$

The *-transform of the Archimedean Property has a different meaning and is true,

$$(\forall x \in R^*)(\exists n \in N^*)x < n.$$

Thus R* is *-Archimedean but not Archimedean.

A second example is the completeness property. R is complete but R* is not. Formally,

$$(\forall x \in \mathcal{P}(R))\phi, \ \neg(\forall x \in \mathcal{P}(R^*))\phi^* ,$$

where ϕ is the formula stating that if $x \neq \emptyset$ and x has an upper bound then x has a least upper bound. However, R* is *-complete, that is

$$(\forall x \in (\mathcal{P}(R))^*)\phi^* .$$

-completeness states that every internal subset of R which is nonempty and has an upper bound has a least upper bound.

A third example is induction on the natural numbers. We have

$$(\forall x \in \mathcal{P}(N))(0 \in x \wedge (\forall y \in x)(y + 1 \in x) \rightarrow x = N),$$

but by considering x = N we see that

$$\neg(\forall x \in \mathcal{P}(N^*))(0 \in x \wedge (\forall y \in x)(y + 1 \in x) \rightarrow x = N^*).$$

However, the hypernatural numbers satisfy *-induction, that is, induction for internal subsets of N*,

$$(\forall x \in (\mathcal{P}(N))^*)(0 \in x \wedge (\forall y \in x)(y + 1 \in x) \rightarrow x = N^*).$$

The following two general consequences of Leibniz' Principle are often useful.

INTERNAL DEFINITION PRINCIPLE

If $\phi(x)$ is an internal formula and c is an internal set, then

$$\{x \in c: \phi(x)\}$$

is an internal set.

Proof Let $b_1,\ldots,b_n$ be the constants in $\phi(x)$, so $\phi(x)$ can be written in the form $\phi(x,b_1,\ldots,b_n)$. For some m, $c,b_1,\ldots,b_n$ all belong to $[V_m(R)]^*$. The following standard sentence is true:

$$(\forall y_1,\ldots,y_n,z \in V_m(R))(\exists u \in V_{m+1}(R))$$
$$u = \{x \in z: \phi(x,y_1,\ldots,y_n)\}.$$

Therefore the *-transform is true, and it follows that

$$\{x \in c: \phi(x)\} \in [V_{m+1}(R)]^* . \quad \dashv$$

THEOREM 2

If $c \in R^*$, X is an internal subset of R^*, and monad(c) $\subseteq$ X, then there is a real $\delta > 0$ such that

$$(c - \delta, c + \delta)^* \subseteq X .$$

In the book of Stroyan, Theorem 2 is called Cauchy's Principle.

Proof We use the fact that R^* is *-complete. The set

$$Y = \{x \in (0,1)^*: (c - x, c + x)^* \subseteq X\}$$

is internal by the Internal Definition Principle. It is nonempty since it contains all positive infinitesimals, and has the upper bound 1. Therefore by *-completeness, Y has a least upper bound x_0. For all $x \approx 0$, $x \leq x_0$, so x_0 is not infinitesimal. Therefore there is a positive real $\delta \leq x_0$, whence $(c - \delta, c + \delta)^* \subseteq X$. $\dashv$

EXAMPLE Let f be an internal function on R^*. Then

$$\{x \in R^*: f \text{ is *-continuous at } x\}$$

is an internal set. Moreover, if f is *-continuous at every $x \approx c$, then f is *-continuous on a neighborhood $(c - \delta, c + \delta)^*$ for some positive real δ.

For an indication of what else can be done with superstructure embeddings we refer to the book of Stroyan. We mention here only that the notion of a *-finite set is extremely useful. A *-finite set is an internal set $b \in V(R^*)$ which satisfies the *-transform of the formula "x is finite". Equivalently, there is an internal one to one mapping of an initial segment of N^* onto b.

For example, it can be easily shown that a real function is Riemann integrable on [a,b] if and only if all the Riemann sums over *-finite partitions of $[a,b]^*$ into infinitesimal subintervals are finite and have the same standard parts. The *-finite sets are likely to be especially important in applications outside mathematics, where it is useful to represent large finite phenomena by *-finite sets (this has been done recently in mathematical economics).

1D* THE ULTRAPOWER CONSTRUCTION

We shall construct a superstructure embedding which satisfies Leibniz' Principle. One way to do this is to apply the compactness theorem for the theory of types (Henkin 1950), as in the book of Robinson. Another method, which we shall adopt here, is the ultrapower construction. Historically, ultrapowers were first applied to the natural numbers by Skolem in 1934. Hewitt studied ultrapowers of the real number field in 1948, and the general construction and fundamental theorem were discovered by Łos in 1955. Since then ultrapowers have been important in model theory.

Actually our construction will have two stages. First we construct an ultrapower of the superstructure V(R). Then we use a method known as the Mostowski collapsing construction to map this ultrapower into a new superstructure V(R*). To simplify our discussion we shall describe a special case of the ultrapower, namely a countably indexed ultrapower. We begin with the definition of an ultrafilter over the set N of natural numbers.

DEFINITION

An <u>ultrafilter</u> U over N is a set of subsets of N such that:

(i) U is closed under supersets; if $X \in U$ and $X \subseteq Y \subseteq N$ then $Y \in U$.

(ii) U is closed under finite intersections; if $X \in U$ and $Y \in U$ then $X \cap Y \in U$.

(iii) $N \in U$ but $\emptyset \notin U$.

(iv) For each $X \subseteq N$, exactly one of the sets X, N - X belongs to U.

An ultrafilter U is said to be <u>free</u> if no finite set belongs to U.

Our construction and results below will apply to ultrafilters over any infinite set I instead of N, but the requirement that U is free must be replaced in the general case by the stronger assumption that some countable intersection of members of U is empty.

ULTRAFILTER THEOREM

There exists a free ultrafilter over N.

<u>Proof</u> By a <u>filter</u> over N we mean a set F of subsets of N which has properties (i) - (iii); that is, F is closed under supersets and finite intersections, and $N \in F$, $\emptyset \notin F$. The set of all cofinite (complements of finite) subsets of N is a filter over N. Let A be the set of all filters F over N such that F contains all cofinite subsets of N. Then A is nonempty and A is closed under unions of chains. By Zorn's Lemma, A has a maximal element U. U is a filter and contains no finite set, because U contains all cofinite sets but $\emptyset \notin U$. To show that U is a free ultrafilter, we prove that if $X \subseteq N$ then there is a filter $V \supseteq U$ which contains either X or N - X, so by maximality, $X \in U$ or $N - X \in U$.

Case 1. For all $Y \in U$, $X \cap Y$ is infinite. X and each $Y \in U$ belongs to the set

$$V = \{Z \subseteq N: Z \supseteq X \cap Y \text{ for some } Y \in U\}.$$

V is a filter over N, because V obviously is closed under supersets and inclusion, and the hypothesis guarantees that each $Z \in V$ is infinite.

Case 2. For some $Y \in U$, $X \cap Y$ is finite. Then for every $W \in U$, $(N - X) \cap W$ is infinite, for otherwise $Y \cap W \in U$ would be finite. Case 1 applies to N - X, so the set

$$V = \{Z \subseteq N: Z \supseteq (N - X) \cap Y \text{ for some } Y \in U\}$$

is a free filter over N with $V \supseteq U$ and $N - X \in V$. ⊣

Hereafter we let U be a free ultrafilter over N.

DEFINITION

A countable sequence $a = \langle a_0, a_1, a_2, \ldots \rangle$ of elements of V(R) is said to be <u>bounded</u> if there is a fixed n such that each a_i belongs to $V_n(R)$.

We shall construct the bounded ultrapower of V(R) modulo U. The elements of the bounded ultrapower are going to be equivalence classes of bounded sequences in V(R), where the equivalence relation is determined by U.

DEFINITION

Two bounded sequences a and b are said to be U-equivalent, $a =_U b$, if

$$\{n \in N: a_n = b_n\} \in U .$$

LEMMA 1 The relation $=_U$ is an equivalence relation on the set of bounded sequences.

Proof $=_U$ is obviously reflexive and symmetric. We show $=_U$ is transitive. Assume $a =_U b$ and $b =_U c$. Let

$$X = \{n: a_n = b_n\},\ Y = \{n: b_n = c_n\},\ Z = \{n: a_n = c_n\} .$$

Then $X \in U$, $Y \in U$, so $X \cap Y \in U$. Also, $X \cap Y \subseteq Z$, so $Z \in U$. Therefore $a =_U c$. ⊣

DEFINITION

The U-equivalence class of a bounded sequence a is denoted by a_U,

$$a_U = \{b: a =_U b\}.$$

The bounded ultrapower $\Pi_U V(R)$ of the set V(R) is the set of all U-equivalence classes of bounded sequences,

$$\Pi_U V(R) = \{a_U: a \text{ is a bounded sequence}\}.$$

The natural embedding of V(R) into $\Pi_U V(R)$ is the mapping

$$i: V(R) \to \Pi_U V(R)$$

defined by

$$i(x) = \langle x,x,x,\ldots\rangle_U$$

for each $x \in V(R)$. That is, $i(X)$ is the U-equivalence class of the constant sequence $\langle x,x,x,\ldots\rangle$.

The <u>relation</u> $\in_U$ on $\Pi_U V(R)$ is defined by

$$a_U \in_U b_U \quad \text{iff} \quad \{n: a_n \in b_n\} \in U.$$

LEMMA 2 The relation $a_U \in_U b_U$ depends only on the equivalence classes a_U and b_U. That is, if $a =_U a'$ and $b =_U b'$ then

$$\{n: a_n \in b_n\} \in U \quad \text{iff} \quad \{n: a'_n \in b'_n\} \in U.$$

<u>Proof</u> Suppose $\{n: a_n \in b_n\} \in U$. Then the set $\{n: a'_n \in b'_n\}$ contains the set

$$\{n: a_n \in b_n\} \cap \{n: a_n = a'_n\} \cap \{n: b_n = b'_n\}.$$

The latter set belongs to U, so

$$\{n: a'_n \in b'_n\} \in U, \ a' =_U b'. \qquad \dashv$$

LEMMA 3 The mapping $i : V(R) \to \Pi_U V(R)$ is one to one.

<u>Proof</u> If $x \neq y$ then $\langle x,x,\ldots\rangle$ and $\langle y,y,\ldots\rangle$ are never equal, and $\emptyset \notin U$, so

$$i(x) \neq i(y). \qquad \dashv$$

The <u>ultrapower</u> of R modulo U is the set of U-equivalence classes of sequences of reals,

$$\Pi_U R = \{b_U: b \text{ maps } N \text{ into } R\}.$$

Thus the following are equivalent:

$$a_U \in \Pi_U R,$$

$$\{n: b_n \in R\} = N \text{ for some } b =_U a,$$

$$\{n: a_n \in R\} \in U,$$

$$a_U \in_U i(R).$$

LEMMA 4 i maps R properly into $\Pi_U R$.

<u>Proof</u> Let a be a one to one mapping of N into R. Then $a_U \in \Pi_U R$. However, for each $r \in R$, $\{n: a_n = r\}$ contains at most one element and hence is not in U, so $a_U \neq i(r)$. ⊣

LEMMA 5 Let $a_U, b_U \in \Pi_U V(R) - \Pi_U R$. If $a_U \neq b_U$ then

$$\{c_U: c_U \in_U a_U\} \neq \{c_U: c_U \in_U b_U\}.$$

<u>Proof</u> Let $a_U \neq b_U$. We may choose the representatives a and b so that $a_n \neq b_n$ for all n. Then one of the sets

$$\{n: a_n - b_n \neq \emptyset\}, \{n: b_n - a_n \neq \emptyset\}$$

belongs to U, because their union is N. Say

$$X = \{n: a_n - b_n \neq \emptyset\} \in U.$$

For each $n \in X$ choose $c_n \in a_n - b_n$, and for $n \notin X$ choose c_n arbitrarily. Then $c_U \in_U a_U$ but $c_U \notin_U b_U$. ⊣

We have defined the ultrapower construction and proved some basic lemmas. Now we carry out the second stage of the construction, the Mostowski collapse.

The ultrapower of R modulo U, hereafter denoted by $R^* = \Pi_U R$, will be the set of atoms for a new superstructure $V(R^*)$. Assume that no element of R^* contains any element of $V(R^*)$.

THEOREM 6

There is a unique one to one mapping j from the bounded ultrapower of V(R) into $V(R^*)$,

$$j: \Pi_U V(R) \to V(R^*),$$

such that

(i) j is the identity function on R^*.
(ii) For all $a_U \in \Pi_U V(R) - R^*$,

$$j(a_U) = \{j(b_U): b_U \in_U a_U\}.$$

j is called the <u>Mostowski collapsing function</u> for $\Pi_U V(R)$.

<u>Proof</u> For each $k \in N$ we consider the ultrapower of $V_k(R)$,

$$\Pi_U V_k(R) = \{a_U: a \text{ maps } N \text{ into } V_k(R)\}.$$

Then

$$\Pi_U V(R) = \bigcup_{k=0}^{\infty} \Pi_U V_k(R),$$

$$\Pi_U R \subseteq \Pi_U V_1(R) \subseteq \Pi_U V_2(R) \subseteq \ldots .$$

We define j restricted to $\Pi_U V_k(R)$ by induction on k. For $k = 0$, $V_0(R) = R$ and j restricted to $\Pi_U R$ is the identity. Suppose we already have j restricted to $\Pi_U V_k(R)$. For $a_U \in \Pi_U V_{k+1}(R) - R^*$ define

$$j(a_U) = \{j(b_U): b_U \in_U a_U\}.$$

This definition is meaningful because whenever $b_U \in_U a_U$ we have

$$\{n: b_n \in V_k(R)\} \in U, \; b_U \in \Pi_U V_k(R),$$

so $j(b_U)$ is already defined.

Conditions (i) - (iii) hold by definition. j is one to one by Lemma 5. To show uniqueness one proves by induction on k that the restriction of j to $\Pi_U V_k(R)$ is the only function with properties (i) - (iii). ⊣

DEFINITION

The embedding $*: V(R) \rightarrow V(R^*)$ is the composition of i and j, that is, for $a \in V(R)$, $a^* = j(i(a))$.

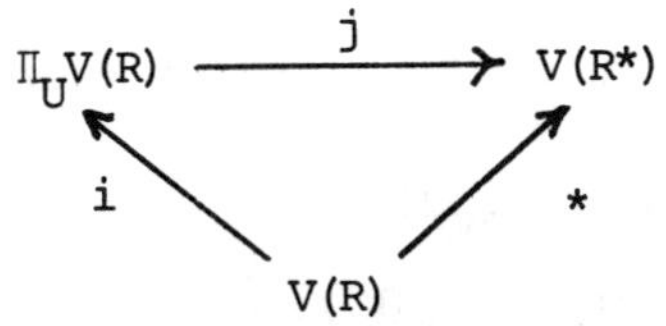

We wish to show that * is a superstructure embedding which satisfies Leibniz' Principle. A minor difficulty is that R is not a subset of R*, but is only mapped one to one into R* by *. We could get around this difficulty by replacing r* by r in R*. Instead we shall slightly change our definition of a superstructure embedding. From now on a superstructure embedding is a one to one mapping of V(R) into V(R*) which maps R onto a proper subset of R* and is such that for $x,y \in V(R)$, $x \in y$ iff $x^* \in y^*$.

LEMMA 7 * is a superstructure embedding.

<u>Proof</u> Since i and j are one to one, * is one to one. i maps R properly into R* and j is the identity on R*, so * maps R properly into R*. We have $R^* = j(i(R)) = {}^*(R)$. Finally, for $x,y \in V(R)$,

$$x \in y \text{ iff } i(x) \in_U i(y) \text{ iff } j(i(x)) \in j(i(y)) \text{ iff } x^* \in y^*. \quad \dashv$$

Leibniz' Principle for * will follow at once from the theorem of Łoś below.

ŁOŚ THEOREM

Let $\phi(v_1,\ldots,v_n)$ be a bounded formula and let $a_{1U},\ldots,a_{nU} \in \Pi_U V(R)$. Then

$$\phi^*(j(a_{1U}),\ldots,j(a_{nU})) \text{ is true}$$

if and only if

$$\{k \in N\colon \phi(a_1(k),\ldots,a_n(k))\} \in U .$$

<u>Proof</u> We use induction on the complexity of formulas. The theorem holds for atomic formulas by definition of $a_U = b_U$ and $a_U \in_U b_U$. For the induction steps it suffices to assume the theorem for $\phi(v_1,\ldots,v_n)$ and $\psi(v_1,\ldots,v_n)$ and prove the theorem for $\neg\phi$, $\phi \wedge \psi$, $(\exists v_1 \in v_2)\phi$, and $(\exists v_1 \in c)\phi$ where $c \in V(R)$.

For $\neg\phi$ we observe that the following are equivalent.

$$\neg\,\phi^*(j(a_{1U}),\ldots,j(a_{nU}))$$

$$\{k \in N\colon \phi(a_1(k),\ldots,a_n(k))\} \notin U$$

$$\{k \in N\colon \neg\phi(a_1(k),\ldots,a_n(k))\} \in U .$$

For $\phi \wedge \psi$ we use the following equivalences.

$$(\phi \wedge \psi)^*(j(a_{1U}),\ldots,j(a_{nU}))$$

$$\phi^*(j(a_{1U}),\ldots) \;\wedge\; \psi^*(j(a_{1U}),\ldots)$$

$$\{k: \phi(a_1(k),\ldots)\} \in U \text{ and } \{k: \psi(a_1(k),\ldots)\} \in U$$

$$\{k: \phi(a_1(k),\ldots)\} \cap \{k: \psi(a_1(k),\ldots)\} \in U$$

$$\{k: \phi \wedge \psi(a_1(k),\ldots)\} \in U .$$

The theorem for $(\exists v_1 \in v_2)\phi$ again is a sequence of equivalent statements.

$$(\exists v_1 \in j(a_{2U}))\phi^*(v_1,j(a_{2U}),\ldots,j(a_{nU}))$$

$$\text{For some } a_{1U} \in_U a_{2U},\ \phi^*(j(a_{1U}),\ldots,j(a_{nU}))$$

$$\{k: \text{For some } a_1(k) \in a_2(k),\ \phi(a_1(k),\ldots,a_n(k))\} \in U$$

$$\{k: (\exists v_1 \in a_2(k))\phi(v_1,a_2(k),\ldots,a_n(k))\} \in U .$$

The proof of the theorem for $(\exists v_1 \in c)\phi$ is similar. ⊣

EXISTENCE THEOREM

There exists a superstructure embedding $*: V(R) \to V(R^*)$ satisfying Leibniz' Principle.

<u>Proof</u> We show that the superstructure embedding * constructed above preserves bounded sentences ϕ. Applying Łoś Theorem to a sentence, we see that ϕ^* is true if and only if $\{k: \phi \text{ is true}\} \in U$. Since $N \in U$ and $\emptyset \notin U$, this is equivalent to the truth of ϕ. ⊣

Since the elements of R^* are now U-equivalence classes of sequences of reals, we are able to give explicit examples of hyperreal numbers with various properties.

EXAMPLES (i) The following are positive infinitesimals.

$$\left\langle 1, \tfrac{1}{2}, \tfrac{1}{3}, \tfrac{1}{4}, \ldots, \tfrac{1}{n}, \ldots \right\rangle_U$$

$$\left\langle 1, \tfrac{1}{2}, \tfrac{1}{4}, \tfrac{1}{8}, \ldots, 2^{-n}, \ldots \right\rangle_U .$$

(ii) The following are positive infinite hyperintegers.

$$\langle 1, 2, 3, 4, \ldots, n, \ldots \rangle_U$$

$$\langle 1, 2, 6, 24, \ldots, n!, \ldots \rangle_U .$$

(iii) The following are infinitely close to but not equal to π.

$$\langle 3, 3.1, 3.14, \ldots, [10^n\pi]/10^n, \ldots \rangle_U$$

$$\langle \pi - 1, \pi - \frac{1}{2}, \pi - \frac{1}{3}, \ldots, \pi - \frac{1}{n}, \ldots \rangle_U .$$

The next theorem verifies these claims.

THEOREM 8

Let $\langle a_1, a_2, a_3, \ldots \rangle$ be a sequence of real numbers and let $r \in R$.

(i) $\langle a_1, a_2, a_3, \ldots \rangle_U \approx r$ for every free ultrafilter U over N if and only if

$$\lim_{n\to\infty} a_n = r .$$

(ii) $\langle a_1, a_2, a_3, \ldots \rangle_U$ is positive infinite for every free ultrafilter U over N if and only if

$$\lim_{n\to\infty} a_n = \infty.$$

<u>Proof</u> We prove (i). First assume

$$\lim_{n\to\infty} a_n = r.$$

Then for each positive real ε, we have $|a_n - r| < \varepsilon$ for all but finitely many $n \in N$, hence

$$\{n: |a_n - r| < \varepsilon\} \in U.$$

By Łoś Theorem, $|a_U -^* r|^* <^* \varepsilon$. Since ε is arbitrary, we have $a_U \approx r$.

Now suppose it is not the case that

$$\lim_{n\to\infty} a_n = r.$$

By the ε,δ condition there is a real $\varepsilon > 0$ such that the set

$$X = \{n: |a_n - r| \geq \varepsilon\}$$

is infinite. Using the proof of the Ultrafilter Theorem one can show that there is a free ultrafilter U over N such that $X \in U$. By Łoś Theorem, $|a_U -^* r|^* \geq^* \varepsilon$, so $a_U \not\approx r$. ⊣

EXAMPLES Here are some sequences a such that the behavior of a_U in R* depends on U.

(i) $<-1, 1, -1, 1, \ldots, (-1)^n, \ldots>_U = \begin{cases} -1 & \text{if } \{n: n \text{ odd}\} \in U \\ 1 & \text{if } \{n: n \text{ even}\} \in U. \end{cases}$

(ii) $<1, \frac{1}{2}, 3, 1, \frac{1}{5}, 6, 1, \frac{1}{7}, 8, \ldots>_U$ is either one, infinitesimal, or infinite depending on U.

(iii) For each $r \in [-1, 1]$ there is a free ultrafilter U over N such that $<\sin 1, \sin 2, \ldots, \sin n, \ldots>_U \approx r$.

THEOREM 9

Let $a: N \rightarrow R$ and $r \in R$.

(i) $<a_1,a_2,a_3,\ldots>_U \approx r$ for some free ultrafilter U over N if and only if a has a subsequence converging to r.

(ii) $<a_1,a_2,a_3,\ldots>_U$ is positive infinite for some free ultrafilter U over N if and only if a has a subsequence diverging to ∞.

Proof Similar to Theorem 8. ⊣

1E* REPRESENTATION AND UNIQUENESS

The hyperreal number system (R,R*,*) is not uniquely characterized by Axioms A-D. In this section we shall state a representation theorem which characterizes all structures which satisfy these axioms, and discuss an additional axiom which does single out a unique structure. The results in this section use deeper methods from model theory and are stated without proof. We shall begin with a theorem which shows that any structure satisfying Axioms A-D can be extended to a superstructure embedding satisfying Leibniz' Principle. Then we shall discuss the representation and uniqueness of superstructure embeddings.

EXTENSION THEOREM

The following are equivalent.

(i) The triple $(R, R^*, *)$ satisfies Axioms A-D for hyperreal numbers.
(ii) There is a superstructure embedding $*_1: V(R) \to V(R^*)$ which satisfies Leibniz' Principle such that $*$ is the elementary part of $*_1$.

We proved the implication (ii) $\to$ (i) in Section 1B*.

We now generalize the notion of a bounded ultrapower. Given a function $f: I \to V(R)$, the <u>equivalence relation</u> of f is the binary relation

$$eq(f) = \{\langle i,j\rangle \in I \times I: f(i) = f(j)\}.$$

DEFINITION

Let U be an ultrafilter over a set I and let F be a filter over the set $I \times I$. The <u>bounded limit ultrapower of</u> $V(R)$ <u>modulo</u> (U,F) is the set

$$\Pi_{U|F} V(R) \subseteq \Pi_U V(R)$$

defined by

$$\Pi_{U|F} V(R) = \{a_U \in \Pi_U V(R): eq(a) \in F\}.$$

We may think of the bounded limit ultrapower as the set of U-equivalence classes of functions which are almost constant modulo F, because $eq(a) \in F$ means that the set of pairs $\langle i,j\rangle$ such that $a(i) = a(j)$ is large enough to be in F.

The <u>bounded limit ultrapower</u> of R modulo (U,F) is the set

$$\Pi_{U|F} R = \{b_U \in \Pi_{U|F} V(R): b \text{ maps } I \text{ into } R\}.$$

We write $R^* = \Pi_{U|F} R$. The <u>natural embedding</u>

$$i: V(R) \to \Pi_{U|F}V(R)$$

and <u>Mostowski collapse</u>

$$j: \Pi_{U|F}V(R) \to V(R^*)$$

are defined exactly as in the case of bounded ultrapowers.

The analogue of Łoś Theorem holds for limit ultrapowers. Usually, * will map R properly into R*. However, a degenerate case can arise where the embedding * is trivial in the following sense.

DEFINITION

An embedding $*: V(R) \to V(R^*)$ is <u>trivial</u> if $R = R^*$ and * is the identity mapping on V(R).

REPRESENTATION THEOREM

The following conditions are equivalent.

(i) $*: V(R) \to V(R^*)$ is either trivial or is a superstructure embedding which satisfies Leibniz' Principle.

(ii) There is a limit ultrapower $\Pi_{U|F}V(R)$ such that * is the composition of the natural embedding i and the Mostowski collapse j,

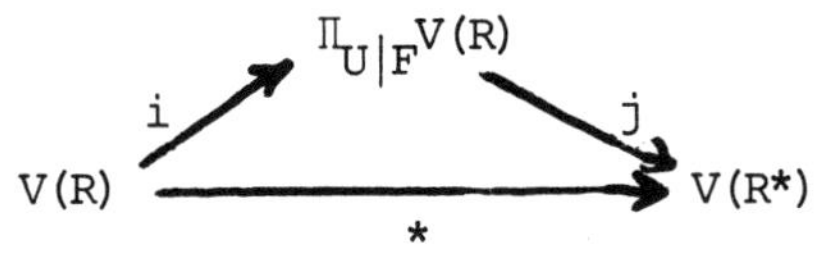

The Representation Theorem shows that the superstructure embeddings $*: V(R) \to V(R^*)$ are precisely the Mostowski collapses of nontrivial bounded limit ultrapowers of V(R).

The real numbers are the unique complete ordered field. By analogy, we would like to uniquely characterize the hyperreal structure (R,R*,*) by some sort of completeness property. However, we run into a set-theoretic difficulty; there are structures R* of arbitrary large cardinal number which satisfy Axioms A-D, so there cannot be a largest one. Two ways around this difficulty are to make R* a proper class rather than a set, or to put a restriction on the cardinal number of R*. We use the second method here because it is simpler. The following axiom has been of central importance in more advanced work in infinitesimal analysis. It is different from the Completeness Property but has a similar appeal.

AXIOM E (Saturation Axiom)

Let S be a set of equations and inequalities involving real functions, hyperreal constants, and variables, such that S has smaller cardinality than R*. If every finite subset of S has a hyperreal solution, then S has a hyperreal solution.

To state a uniqueness theorem we need the notion of an inaccessible cardinal.

DEFINITION

A cardinal number κ is said to be inaccessible if:

(i) For any set x of cardinality less than κ, the power set $\mathcal{P}(x)$ has cardinality less than κ.

(ii) For any set of sets $X = \{x_i : i \in I\}$ such that I and each x_i has cardinality less than κ, the union $\bigcup_{i\in I} x_i$ has cardinality less than κ.

The axiom of inaccessibility in set theory is the axiom that there exists an uncountable inaccessible cardinal number.

UNIQUENESS THEOREM

There is up to isomorphism a unique structure (R,R*,*) such that Axioms A-E are satisfied and the cardinality of R* is the first uncountable inaccessible cardinal number.

There is a similar uniqueness theorem for superstructure embeddings.

DEFINITION

A superstructure embedding $*: V(R) \to V(R^*)$ is said to be saturated if for every set X of internal sets such that the cardinality of X is less than the cardinality of R*, if every finite subset of X has nonempty intersection then X has nonempty intersection.

UNIQUENESS THEOREM FOR SUPERSTRUCTURE EMBEDDINGS

There is up to isomorphism a unique superstructure embedding $*: V(R) \to V(R^*)$ such that:

(i) * satisfies Leibniz' Principle;

(ii) * is saturated;

(iii) Both R* and the set of all internal sets have cardinality equal to the first uncountable inaccessible cardinal.

The results of this section are all proved, in a somewhat different form, in the book of Chang and Keisler. The Extension and Representation Theorems are due to the author, and the Uniqueness Theorem is due to Morley and Vaught.

2 DIFFERENTIATION

Throughout this monograph it will be understood that f,g,... denote real functions rather than hyperreal functions. We shall use the hyperreal numbers as a tool to define the notions of limit, derivative, continuity, and integral for real functions. The reader is warned that the notion of continuity at a point splits into two different notions when applied to hyperreal functions. These notions, called S-continuity and *-continuity, are discussed in the optional Section 1C* but will not be needed hereafter.

Section 2B contains a rigorous treatment of infinitesimal microscopes and telescopes suggested by Keith Stroyan. The remaining material in this chapter is also given in Elementary Calculus but is repeated here to make this monograph complete.

2A DERIVATIVES (§2.1,§2.2)

DEFINITION

S is said to be the slope of a real function f at a real point a if

$$S = st\left(\frac{f(a + \Delta x) - f(a)}{\Delta x}\right)$$

for every nonzero infinitesimal Δx.

The derivative of a real function f is the real function f' such that:

$$f'(x) = \text{slope of } f \text{ at } x \text{ if it exists,}$$

$$f'(x) \text{ is undefined otherwise.}$$

We shall show in Chapter 5 that this definition is equivalent to the standard definition of derivative.

We say that f is <u>differentiable</u> at a if the slope of f at a exists.

REMARK We see from the definition that f is differentiable at a if and only if

(1) $f(x)$ is defined for all $x \approx a$,

(2) The quotient $\frac{f(a + \Delta x) - f(a)}{\Delta x}$ is finite and has the same standard part for all nonzero $\Delta x \approx 0$.

Thus by Corollary 4 in Section 1D, if f is differentiable at a then $f(x)$ is defined for all real x in some neighborhood of a.

To take full advantage of the Leibniz notation we use independent and dependent variables. We first make these notions precise. If we are given a system of formulas which has the same solution set as a simple equation $y = f(x)$, we say that y is a <u>function</u> of x, or that y <u>depends</u> on x, and we call x the <u>independent variable</u> and y the <u>dependent variable</u>. When y depends on x, we introduce one new independent variable Δx and two new dependent variables Δy and dy. Δy is called the <u>increment</u> of y and its dependence on x and Δx is given by the equation

$$\Delta y = f(x + \Delta x) - f(x).$$

Thus when $f'(x)$ exists its value is

$$f'(x) = st\left(\frac{\Delta y}{\Delta x}\right).$$

We call dy the <u>differential</u> of y, and its dependence on x and Δx is given by

$$dy = f'(x)\Delta x,$$

with the understanding that dy exists only when $f'(x)$ exists. As usual we write

$$dx = \Delta x,$$

so we have the familiar equations

$$dy = f'(x)dx, \quad f'(x) = \frac{dy}{dx}.$$

Geometrically, we define the <u>tangent line</u> to the curve $y = f(x)$ at a point (a,b) on the curve to be the line through (a,b) with slope $f'(a)$. Thus

$$\Delta y = \text{change in } y \text{ along curve}$$

$$dy = \text{change in } y \text{ along tangent line.}$$

Usually we shall be interested in the case where Δx is infinitesimal. Remember that the increment Δy and differential dy are dependent variables which depend on x and Δx. The next theorem shows the relationship between dy and Δy when $\Delta x \approx 0$.

DEFINITION

Let Δx be a nonzero infinitesimal. We say that u and v are <u>infinitely close compared to</u> Δx, in symbols

$$u \approx v \quad (\text{compared to } \Delta x),$$

if $\dfrac{u}{\Delta x} \approx \dfrac{v}{\Delta x}$.

The relation $\approx$ (compared to Δx) is obviously an equivalence relation. It is closely related to the large and small oh notation. We shall explain that notation now but will not use it in this book. For each nonzero infinitesimal Δx, we define

$$O(\Delta x) = \{u: u/\Delta x \text{ is finite}\},$$

$$o(\Delta x) = \{u: u/\Delta x \text{ is infinitesimal}\}.$$

Both $O(\Delta x)$ and $o(\Delta x)$ are ideals in galaxy(0), and

$$o(\Delta x) \subseteq O(\Delta x) \subseteq \text{monad}(0).$$

We see from the definitions that

$$u \approx v \text{ (compared to } \Delta x) \quad \text{iff} \quad u - v \in o(\Delta x).$$

INCREMENT THEOREM

Let $y = f(x)$, and suppose $f'(x)$ exists and Δx is a nonzero infinitesimal. Then

$$\Delta y = f'(x)\Delta x + \varepsilon\Delta x = dy + \varepsilon\Delta x$$

for some infinitesimal ε. In other words,

$$\Delta y \approx dy \text{ (compared to } \Delta x).$$

Proof Take

$$\varepsilon = \frac{\Delta y}{\Delta x} - f'(x).$$

Then $\varepsilon \approx 0$. Multiplying by Δx,

$$\varepsilon\Delta x = \Delta y - f'(x)\Delta x,$$

$$\Delta y = f'(x)\Delta x + \varepsilon\Delta x.$$

⊣

2B INFINITESIMAL MICROSCOPES AND INFINITE TELESCOPES

In Elementary Calculus we frequently used the pictorial devices of infinitesimal microscopes and infinite telescopes to illustrate definitions and theorems about hyperreal numbers. The following precise definition was suggested by Keith Stroyan and may help the instructor to consistently apply the device in new situations.

Given a point $P(a,b)$ in the hyperreal plane and a positive hyperreal number ε, the ε-disc about P is defined as the set of all points (x,y) at distance at most ε from P, that is

$$(x - a)^2 + (y - b)^2 \le \varepsilon^2.$$

DEFINITION

Let $P(a,b)$ be a point in the hyperreal plane and δ be a positive infinitesimal. The δ-infinitesimal microscope aimed at P is the mapping m from the 2δ-disc about P onto the 2-disc about the origin given by

$$m(a + \delta x,\ b + \delta y) = (x,y),\quad x^2 + y^2 \le 4.$$

Thus m maps (a,b) to $(0,0)$, magnifies distances by $1/\delta$, and preserves directions.

The 2δ-disc about P is called the field of view of the microscope. A drawing of a δ-infinitesimal microscope will distinguish two points if and only if the distance between them is not infinitesimal compared to δ. Thus a point (x,y) is infinitely close to (a,b) if and only if it is in the field of view of some infinitesimal microscope aimed at (a,b). As an example we shall discuss the use of infinitesimal microscopes in giving a picture of the slope of a function.

By the Increment Theorem, for each real point a we have $f'(a) = S$ if and only if for every nonzero infinitesimal Δx, the curve at $a + \Delta x$ is infinitely close to the tangent line at $a + \Delta x$ compared to Δx, that is,

$$f(a + \Delta x) \approx f(a) + S\Delta x \quad (\text{compared to } \Delta x).$$

It follows that $f'(a) = S$ at a real point a if and only if the curve $y = f(x)$ looks like a straight line with slope S in the field of any infinitesimal microscope aimed at $(a,f(a))$. This is illustrated in Figure 2a.

One would also like to use infinitesimal microscopes to illustrate the difference between the tangent line and the curve when the slope exists. In Elementary Calculus we simply used artistic license and drew a picture like Figure 2b with the curvature exaggerated. A more sophisticated but accurate picture can be drawn by using a more powerful $(\Delta x)^2$-infinitesimal microscope within a (Δx)-infinitesimal microscope, as in Figure 2c. The curve and tangent line are indistinguishable in the (Δx)-infinitesimal microscope but can usually be distinguished in the $(\Delta x)^2$-infinitesimal microscope aimed at $(a + \Delta x,\ f(a + \Delta x))$.

We now turn to the notion of an infinite telescope.

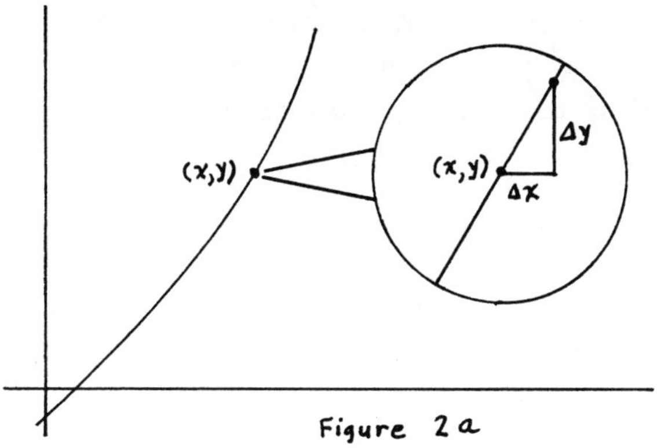

Figure 2a

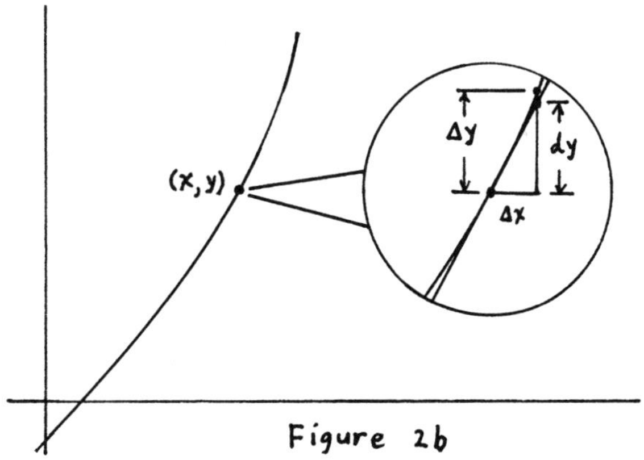

Figure 2b

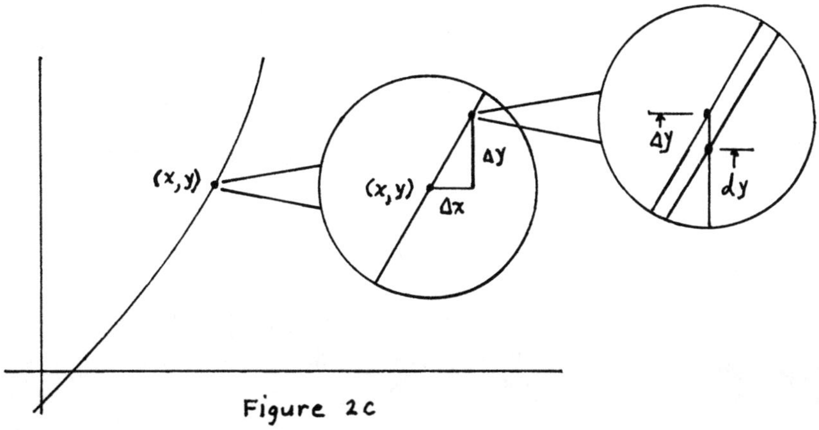

Figure 2c

DEFINITION

Let (a,b) be a point in the hyperreal plane infinitely far from the origin, that is, such that $a^2 + b^2$ is infinite. By the <u>infinite telescope aimed at</u> (a,b) we mean the mapping t of the 2-disc about (a,b) onto the 2-disc about the origin given by

$$t(a + x, b + y) = (x,y).$$

Thus t simply translates the 2-disc about (a,b) to the 2-disc about the origin, and preserves distances and directions. In <u>Elementary Calculus</u>, limits of the form

$$\lim_{x\to\infty} f(x) = L$$

are illustrated with an infinitesimal microscope within an infinite telescope aimed at a point (H,L), H positive infinite. See Chapter 5 for the hyperreal definition of infinite limits.

2C PROPERTIES OF DERIVATIVES (§2.3,§2.4)

The familiar rules for derivatives can be obtained quite easily from the rules for standard parts in Chapter 1.

Given a term τ, we shall write $d\tau$ for dy, and $d\tau/dx$ for dy/dx, where y is the dependent variable given by $y = \tau$. This often saves space. For example, we can write $d(u + v)$ without introducing a new variable $y = u + v$.

THEOREM 1

Suppose u and v depend on the independent variable x. Then for any real value of x where du/dx and dv/dx exist we have

(i) (Sum Rule) $\dfrac{d(u + v)}{dx} = \dfrac{du}{dx} + \dfrac{dv}{dx}$

(ii) (Constant Rule) For any real number c, $\dfrac{d(cu)}{dx} = c\,\dfrac{du}{dx}$

(iii) (Product Rule) $\dfrac{d(uv)}{dx} = u\,\dfrac{dv}{dx} + v\,\dfrac{du}{dx}$

(iv) (Quotient Rule) If $v \neq 0$, $\dfrac{d(u/v)}{dx} = \dfrac{v\,\dfrac{du}{dx} - u\,\dfrac{dv}{dx}}{v^2}$.

Proof For each part we let Δx be a nonzero infinitesimal.

Sum Rule: Let $y = u + v$. Then

$$\Delta y = (u + \Delta u) + (v + \Delta v) - (u + v) = \Delta u + \Delta v,$$

$$\frac{\Delta y}{\Delta x} = \frac{\Delta u}{\Delta x} + \frac{\Delta v}{\Delta x},$$

$$st\left(\frac{\Delta y}{\Delta x}\right) = st\left(\frac{\Delta u}{\Delta x} + \frac{\Delta v}{\Delta x}\right) = st\left(\frac{\Delta u}{\Delta x}\right) + st\left(\frac{\Delta v}{\Delta x}\right),$$

$$\frac{dy}{dx} = \frac{du}{dx} + \frac{dv}{dx}.$$

Constant Rule: Let $y = cu$. Then

$$\Delta y = c(u + \Delta u) - cu = c\Delta u,$$

$$\frac{\Delta y}{\Delta x} = c\,\frac{\Delta u}{\Delta x},$$

$$st\left(\frac{\Delta y}{\Delta x}\right) = st\left(c\,\frac{\Delta u}{\Delta x}\right) = c\ st\left(\frac{\Delta u}{\Delta x}\right),$$

$$\frac{dy}{dx} = c\,\frac{du}{dx}.$$

Product Rule: Let $y = uv$.

$$\Delta y = (u + \Delta u)(v + \Delta v) - uv$$
$$= u\Delta v + v\Delta u + \Delta u\Delta v,$$

$$\frac{\Delta y}{\Delta x} = u\,\frac{\Delta v}{\Delta x} + v\,\frac{\Delta u}{\Delta x} + \Delta u\,\frac{\Delta v}{\Delta x},$$

$$st\left(\frac{\Delta y}{\Delta x}\right) = st\left(u\,\frac{\Delta v}{\Delta x} + v\,\frac{\Delta u}{\Delta x} + \Delta u\,\frac{\Delta v}{\Delta x}\right)$$
$$= u\ st\left(\frac{\Delta v}{\Delta x}\right) + v\ st\left(\frac{\Delta u}{\Delta x}\right) + 0\ st\left(\frac{\Delta v}{\Delta x}\right)$$

$$\frac{dy}{dx} = u\,\frac{dv}{dx} + v\,\frac{du}{dx}.$$

Quotient Rule: Let $y = \frac{u}{v}$, $v \neq 0$.

$$\Delta y = \frac{u + \Delta u}{v + \Delta v} - \frac{u}{v} = \frac{(u + \Delta u)v - (v + \Delta v)u}{v(v + \Delta v)} =$$

$$= \frac{v\Delta u - u\Delta v}{v(v + \Delta v)}$$

$$\frac{\Delta y}{\Delta x} = \frac{v\,\frac{\Delta u}{\Delta x} - u\,\frac{\Delta v}{\Delta x}}{v(v + \Delta v)}$$

$$st\left(\frac{\Delta y}{\Delta x}\right) = \frac{v\ st\left(\frac{\Delta u}{\Delta x}\right) - u\ st\left(\frac{\Delta v}{\Delta x}\right)}{v^2}$$

$$\frac{dy}{dx} = \frac{v\,\frac{du}{dx} - u\,\frac{dv}{dx}}{v^2}\ .$$

⊣

THEOREM 2 (Power Rule)

If x is a positive real number and r is any rational number, then

$$\frac{d(x^r)}{dx} = rx^{r-1}\ .$$

Proof Case 1. r is a positive integer. The proof is an easy induction using the Product Rule.

Case 2. $r = 1/n$ for some positive integer n. Let $y = x^{1/n}$ and let Δx be a nonzero infinitesimal. Consider

$$\Delta y = (x + \Delta x)^{1/n} - x^{1/n}\ .$$

$\Delta y \neq 0$ because $x + \Delta x \neq x$. Δy is infinitesimal because

$$st(\Delta y) = st((x + \Delta x)^{1/n}) - st(x^{1/n})$$
$$= x^{1/n} - x^{1/n} = 0.$$

Now

$$x = y^n,$$

$$\frac{dx}{dy} = ny^{n-1},$$

$$\frac{\Delta x}{\Delta y} \approx ny^{n-1}\ .$$

Therefore

$$\frac{\Delta y}{\Delta x} \approx \frac{1}{ny^{n-1}} = \frac{1}{n}\,x^{(\frac{1}{n})-1},$$

$$\frac{dy}{dx} = \frac{1}{n} x^{(\frac{1}{n})-1} .$$

Case 3. r is a positive rational. This follows from Cases 1 and 2 using the fact that

$$x^{m/n} = (x^{1/n})^m .$$

Case 4. r is a negative rational. This follows from Case 3 using the Quotient Rule. ⊣

If $r = m/n$ where n is odd, the above Power Rule also holds for negative values of x. If $r = m/n$ where n is even, then x^r is undefined in the real number system.

2D CHAIN RULE (§2.5,§2.6)

The Chain Rule can be proved in a natural way using the Increment Theorem.

CHAIN RULE

Let f and G be real functions and let g be the composition

$$g(t) = G(f(t)).$$

For any real value of t where $f'(t)$ and $G'(f(t))$ exist, $g'(t)$ also exists and

$$g'(t) = G'(f(t))f'(t).$$

Proof Let $x = f(t)$, $y = g(t) = G(x)$. Let $\Delta t \neq 0$ be infinitesimal. By the Increment Theorem for $x = f(t)$, Δx is infinitesimal. By the Increment Theorem for $y = G(x)$,

$$\Delta y = G'(x)\Delta x + \varepsilon\Delta x$$

for some infinitesimal ε. Dividing by Δt and taking standard parts,

$$\frac{\Delta y}{\Delta t} = G'(x)\,\frac{\Delta x}{\Delta t} + \varepsilon\,\frac{\Delta x}{\Delta t}\,,$$

$$\frac{dy}{dt} = G'(x)\,\frac{dx}{dt} + 0\,,$$

$$g'(t) = G'(f(t))f'(t)\,. \qquad \dashv$$

The Chain Rule with dependent variables is stated in the following form.

Let $x = f(t)$, $y = g(t) = G(x)$, and suppose $g'(t)$ and $G'(x)$ exist. Then

$$\frac{dy}{dt} = \frac{dy}{dx}\,\frac{dx}{dt}$$

where dx/dt and dy/dt are computed with t as the independent variable, and dy/dx is computed with x as the independent variable.

As in the standard calculus treatment, the Chain Rule is not trivial because dy has one meaning when x is the independent variable and a different meaning when t is the independent variable,

$$dy = G'(x)dx \text{ and } dy = g'(t)dt.$$

Higher derivatives are defined in the usual way. Thus

$$f'' \text{ is the derivative of } f',$$

$$f^{(n+1)} \text{ is the derivative of } f^{(n)}\,.$$

The nth <u>differential</u> of y can be considered separately and is defined by

$$d^n y = f^{(n)}(x)dx^n$$

where dx^n means $(dx)^n$. Notice that when dx is infinitesimal, dx^2 is a much smaller infinitesimal and d^2y is the product of the real number $f''(x)$ and dx^2.

CONTINUOUS FUNCTIONS 3

In this chapter we go considerably beyond the treatment given in Chapter 3 of Elementary Calculus. Several proofs which were omitted or only sketched there are given fully here.

3A LIMITS AND CONTINUITY (§3.1,§3.2)

We shall now use hyperreal numbers to define the notions of a limit and of a continuous function. In Chapter 5 we show that these definitions are equivalent to the standard definitions.

DEFINITION

Let L and c be real numbers. L is the limit of f(x) as x approaches c, in symbols

$$L = \lim_{x \to c} f(x),$$

if whenever $x \approx c$ but $x \neq c$, we have $f(x) \approx L$. If there is no such L we say that the limit does not exist.

In Elementary Calculus we made the intuitive statement that when $\lim_{x \to c} f(x) = L$, we can see the entire part of the hyperreal graph of f(x),

$x \underset{\neq}{\approx} c$, in an infinitesimal microscope aimed at (c,L). This is a simplification which does not conform to the precise definition of infinitesimal microscope given in Section 2B, because an ε-infinitesimal microscope only has a field of radius 2ε. A more exact statement is as follows. $\lim_{x \to c} f(x) = L$ if and only if for every $x \underset{\neq}{\approx} c$ the point (x,f(x)) belongs to the union of the fields of all infinitesimal microscopes aimed at (c,L).

Limits can often be formed by computing standard parts. We see from the definition that if st(f(x)) = L for all x infinitely close but not equal to c, then

$$\lim_{x \to c} f(x) = L.$$

Notice that if $\lim_{x \to c} f(x)$ exists then f(x) must be defined for all $x \neq c$ in the monad of c. It follows from Corollary 4 in Section 1D that f(x) must be defined for all real $x \neq c$ in some neighborhood of c. The next theorem is immediate.

THEOREM 1

The slope of f at a is given by the limit

$$f'(a) = \lim_{\Delta x \to 0} \frac{f(a + \Delta x) - f(a)}{\Delta x}.$$

The limit with respect to a set $Y \subseteq R$ is defined as follows.

DEFINITION

Let L and c be real numbers. L is the <u>limit</u> of f(x) as x approaches c in Y,

$$L = \lim_{\substack{x \to c \\ x \in Y}} f(x),$$

if whenever $x \in Y^*$ and $x \approx c$ but $x \neq c$, we have $f(x) \approx L$.

Important special cases are the one-sided limits, defined by

$$\lim_{x\to c^-} f(x) = \lim_{\substack{x\to c\\ x<c}} f(x) ,$$

$$\lim_{x\to c^+} f(x) = \lim_{\substack{x\to c\\ x>c}} f(x) .$$

The following result is easy.

THEOREM 2

$\lim_{x\to c} f(x)$ exists if and only if both one-sided limits exist and are equal,

$$\lim_{x\to c^-} f(x) = \lim_{x\to c^+} f(x) .$$

The rules for standard parts lead at once to the familiar rules for limits.

THEOREM 3 (Rules for Limits)

Suppose the limits

$$\lim_{x\to c} f(x),\ \lim_{x\to c} g(x)$$

both exist.

(i) For any constant k, $\lim_{x\to c} k\, f(x) = k \lim_{x\to c} f(x)$.

(ii) $\lim_{x\to c} (f(x) + g(x)) = \lim_{x\to c} f(x) + \lim_{x\to c} g(x)$.

(iii) $\lim_{x\to c} f(x)g(x) = (\lim_{x\to c} f(x))(\lim_{x\to c} g(x))$.

(iv) If $\lim_{x\to c} g(x) \neq 0$, $\lim_{x\to c} (f(x)/g(x)) = (\lim_{x\to c} f(x))/(\lim_{x\to c} g(x))$.

<u>Proof</u> To illustrate we prove (ii). Let $x \approx c$ but $x \neq c$. Then

$$\lim_{x\to c} (f(x) + g(x)) = st(f(x) + g(x))$$

$$= st(f(x)) + st(g(x)) = \lim_{x\to c} f(x) + \lim_{x\to c} g(x). \qquad \dashv$$

DEFINITION

f is <u>continuous</u> at a real point c if f(c) exists and $\lim_{x \to c} f(x) = f(c)$.

Going back to first principles we see that f is continuous at c if and only if whenever $x \approx c$, $f(x) \approx f(c)$. Notice that if f is continuous at c then f(x) is defined for all real x in some neighborhood of c. It follows from Theorem 3 that sums, products, and quotients of continuous functions are continuous, provided the denominator is not 0.

THEOREM 4

If f is differentiable at c then f is continuous at c.

<u>Proof</u> Let f be differentiable at c. Then f(c) exists. Let $x \approx c$ but $x \neq c$. Then

$$\frac{f(x) - f(c)}{x - c}$$

is finite and $x - c$ is infinitesimal. It follows that $f(x) - f(c)$ is infinitesimal, so $f(x) \approx f(c)$. ⊣

THEOREM 5

Compositions of continuous functions are continuous. That is, if f is continuous at c, and G is continuous at f(c), then $g(x) = G(f(x))$ is continuous at c.

<u>Proof</u> Let $x \approx c$. Then $f(x) \approx f(c)$, so

$$g(x) = G(f(x)) \approx G(f(c)) = g(c). \qquad ⊣$$

We now define continuity and uniform continuity on a set Y of real numbers.

DEFINITION

Let Y be a subset of the domain of f. f is <u>continuous</u> on Y if whenever $c \in Y$, $x \approx c$, and $x \in Y^*$, we have $f(x) \approx f(c)$.

f is uniformly continuous on Y if whenever $x, y \in Y^*$ and $x \approx y$, we have $f(x) \approx f(y)$.

Obviously, uniform continuity implies continuity.

A set Y of reals is said to be compact if it is closed and bounded. For example, each closed interval $[a,b]$ is compact. From Section 1C we see that Y is compact if and only if for every $y \in Y^*$, y is finite and $st(y) \in Y$.

THEOREM 6

Let Y be a compact set of reals. If f is continuous on Y then f is uniformly continuous on Y.

Proof Suppose f is continuous on Y. Let $x,y \in Y^*$ and $x \approx y$. Since Y is compact, x is finite and $c \in Y$ where $c = st(x)$. Since $x \approx y$, $c = st(y)$. By continuity,

$$f(x) \approx f(c), \quad f(y) \approx f(c).$$

Thus $f(x) \approx f(y)$, and f is uniformly continuous on Y. ⊣

If f is uniformly continuous on a set Y then f is obviously uniformly continuous on any subset of Y. The following theorem allows us to extend the domain of a uniformly continuous function.

THEOREM 7

(i) Let f be uniformly continuous on an interval I. Then there is a function g which agrees with f on I and is uniformly continuous on the whole real line.

(ii) Suppose the derivative f' of f is uniformly continuous on an interval I. Then there is a function g which agrees with f on I such that g' is uniformly continuous on the whole real line.

Proof We give the proof for the case where I is an open interval of the form $(-\infty,b)$. Other cases are similar.

(i) We first show that $\lim_{x \to b^-} f(x)$ exists. Let $x < b$ and $x \approx b$. Assume for the moment that $f(x)$ is infinite. Then every real

$u < b$ is a partial hyperreal solution of

$$u < y < b, \ |f(u) - f(y)| \geq 1. \tag{1}$$

Let $u_1 < b$, $u_1 \approx b$. By the Partial Solution Theorem there is a hyperreal y_1 such that (1) holds. But then $u_1 \approx y_1$, $f(u_1) \not\approx f(y_1)$, contradicting the uniform continuity of f. Therefore f(x) is finite and has a standard part B. For all $y < b$ with $y \approx x$ we have $f(y) \approx f(x)$ and hence $f(y) \approx B$, so $B = \lim_{x \to b^-} f(x)$. Now let g be the function

$$g(x) = \begin{cases} f(x) & \text{if } x < b \\ B & \text{if } x \geq b\ . \end{cases}$$

Then g agrees with f on $(-\infty,b)$ and is uniformly continuous on the whole real line.

(ii) From the proof of (i), the limits

$$B = \lim_{x \to b^-} f(x), \quad C = \lim_{x \to b^-} f'(x)$$

both exist. The function

$$g(x) = \begin{cases} f(x) & \text{if } x < b \\ B + C(x - b) & \text{if } x \geq b \end{cases}$$

agrees with f on $(-\infty,b)$ and has a uniformly continuous derivative on the whole real line. ⊣

3B HYPERINTEGERS (§3.3)

Hyperintegers are a basic tool in several areas of the calculus, including integration and infinite series. In this chapter they are used in the proofs of the Intermediate and Extreme Value Theorems.

DEFINITION

The set of integers is denoted by Z. The natural extension Z* of Z is called the set of <u>hyperintegers</u>.

Note that since $Z^* \cap R = Z$, a real number is an integer if and only if it is a hyperinteger.

THEOREM 1

(i) Z^* is a subring of R^*. That is, sums, differences, and products of hyperintegers are hyperintegers.

(ii) For every hyperreal number y there exists a hyperinteger K such that $K \le y < K + 1$.

(iii) There are positive infinite and negative infinite hyperintegers.

(iv) For each hyperinteger K, K + 1 is the smallest hyperinteger greater than K.

(v) Every finite hyperinteger is an integer, that is,

$$Z^* \cap \text{galaxy}(0) = Z.$$

Proof (i) Every real solution of

$$x \in Z, \ y \in Z \tag{1}$$

is a solution of

$$x + y \in Z. \tag{2}$$

Hence every hyperreal solution of (1) is a solution of (2), and the sum of two hyperintegers is a hyperinteger. The proofs for differences and products are similar.

(ii) For each real number y there is a unique integer $n = [y]$, called the integral part of y, such that

$$n \le y < n + 1.$$

Thus the system of formulas

$$[y] \le y, \ y < [y] + 1, \ [y] \in Z \tag{3}$$

holds for all real numbers y. Therefore (3) holds for all hyperreal numbers, and (ii) follows with $K = [y]$.

(iii) Let H be a positive infinite hyperreal number. Then $K = [H] + 1$ is a hyperinteger which is greater than H and hence positive infinite.

(iv) Given an integer n, n + 1 is the smallest integer greater than n. Therefore every real solution of

$$n \in Z, \; n < x < n + 1 \tag{4}$$

is a solution of

$$x \notin Z, \; n + 1 \in Z. \tag{5}$$

Thus every hyperreal solution of (4) is a solution of (5), and (iv) follows.

(v) Let K be a finite hyperinteger. Then st(K) is real, so there is an integer n with

$$n \le st(K) < n + 1.$$

It follows that

$$0 \le |n - K| < 1.$$

However, $|n - K|$ is a hyperinteger by (i), so by (iv) we must have $|n - K| = 0$, $n = K$, $K \in Z$. ⊣

A frequent construction in the calculus is the partition of a closed interval [a,b] into infinitely many subintervals of equal infinitesimal length. When x and y are hyperreal numbers, we shall call the set

$$[x,y]^* = \{z \in R^*: x \le z \le y\}$$

a <u>hyperreal closed interval</u>. If $a \le x \le y \le b$ we call $[x,y]^*$ a <u>hyperreal subinterval</u> of $[a,b]^*$. When $0 < \Delta x$ and $\Delta x \approx 0$, we call $[x,x + \Delta x]^*$ an <u>infinitesimal interval</u>. Since no ambiguity can arise, we sometimes drop the star on an infinitesimal interval, writing $[x,x + \Delta x]$ for $[x,x + \Delta x]^*$. In <u>Elementary Calculus</u> we always wrote $[x,x + \Delta x]$ instead of $[x,x + \Delta x]^*$. Given a hyperinteger $H > 0$, the closed hyperreal interval $[a,b]^*$ may be partitioned into subintervals of length $\delta = (b - a)/H$. The partition points are

$$a, \; a + \delta, \; a + 2\delta, \; \ldots, \; a + K\delta, \; \ldots, \; a + H\delta = b$$

where K runs over the hyperintegers from 0 to H. If H is infinite, each subinterval will have infinitesimal length $\delta = (b - a)/H$, and the partition is called an <u>infinite partition</u> of $[a,b]^*$, or sometimes an <u>infinite partition</u> of [a,b]. A typical subinterval has the form

$$[a + K\delta, \; a + (K + 1)\delta]^*.$$

COROLLARY 2 Given a closed real interval [a,b] and a positive hyperinteger H, let $\delta = (b - a)/H$. Then $[a,b]^*$ is the union of the subintervals

$$[a + K\delta, a + (K + 1)\delta)^*$$

where $K \in Z^*$ and $0 \leq K < H$.

Proof Let $x \in [a,b]^*$. By Theorem 1 there is a $K \in Z^*$ such that

$$K \leq \frac{x - a}{\delta} < K + 1 .$$

It follows that

$$0 \leq K < \frac{b - a}{\delta} = H, \quad a + K\delta \leq x < a + (K + 1)\delta.$$ ⊣

3C CONTINUOUS FUNCTIONS ON AN INTERVAL (§3.4 - §3.9)

We use hyperintegers to prove the Intermediate and Extreme Value Theorems.

INTERMEDIATE VALUE THEOREM

Suppose the real function f is continuous on the closed interval [a,b]. Then for every real number D between f(a) and f(b) there is a point $c \in [a,b]$ such that $f(c) = D$.

Proof We may assume $f(a) \leq D \leq f(b)$. Consider a positive integer n and the finite partition

$$a, a + \delta, a + 2\delta, \ldots, a + b$$

where $\delta = (b - a)/n$. The value of f must cross D on one of the subintervals, so there is an integer m such that

$$0 \leq m < n, \quad f(a + m\delta) < D \leq f(a + (m + 1)\delta). \qquad (1)$$

Thus every real solution of the system of formulas

$$n \in Z, \quad 0 < n, \quad \delta = (b - a)/n \qquad (2)$$

is a partial real solution of (1). Now let n_1 be a positive infinite hyperinteger and let $\delta_1 = (b - a)/n_1$. By the Partial Solution Theorem there exists m_1 such that (1) holds. Let $c = st(a + m_1\delta_1)$. We show that $a \le c \le b$ and $f(c) = D$. We have

$$a \le a + m_1\delta_1, \quad a + (m_1 + 1)\delta_1 \le a + n_1\delta_1 = b.$$

Taking standard parts, $a \le c \le b$. Since f is continuous on [a,b],

$$f(c) = st(f(a + m_1\delta_1)) \le D,$$

$$f(c) = st(f(a + (m_1 + 1)\delta_1)) \ge D.$$

Therefore $f(c) = D$. ⊣

DEFINITION

f has a *maximum* at c if $f(c) \ge f(x)$ for all x in the domain of f. A *minimum* of f is defined analogously.

THEOREM 1

Suppose $c \in R$ and f has a maximum at c, that is, $f(c) \ge f(x)$ for all real x in the domain of f. Then c is also a maximum of f with respect to the hyperreal numbers, that is, $f(c) \ge f(x)$ for all hyperreal x in the domain of f*.

A similar result holds for a minimum.

Proof The formulas $f(x) = f(x)$ and $f(c) \ge f(x)$ have the same real solutions, and hence the same hyperreal solutions. ⊣

DEFINITION

f has a *local maximum* at c if c has a real neighborhood $(c - r, c + r)$ such that $f(c) \le f(x)$ (hence f(x) is defined) for all $x \in (c - r, c + r)$. Local minima are defined in a similar way.

THEOREM 2

Let $c \in R$. Then f has a local maximum at c if and only if $f(c) \ge f(x)$ for all hyperreal $x \approx c$.

Proof Suppose f has a local maximum at c, so $f(c) \geq f(x)$ for all $x \in (c - r, c + r)$. By Theorem 1, $f(c) \geq f(x)$ for all hyperreal x such that $c - r < x < c + r$. Now suppose f does not have a local maximum at c. Case 1: f is not defined on a neighborhood of c. Then $f(x)$ is undefined for some $x \approx c$. Case 2: f is defined on some neighborhood of c. Then for each positive real r there is a point $x \in (c - r, c + r)$ such that $f(c) < f(x)$. Thus every real $r > 0$ is a partial real solution of

$$c - r < x < c + r, \; f(c) < f(x). \tag{3}$$

Let r_1 be positive infinitesimal. Then r_1 is a partial hyperreal solution of (3). Hence there exists $x_1 \approx c$ with $f(c) < f(x_1)$. ⊣

EXTREME VALUE THEOREM

Let f be continuous on the closed interval [a,b], and let [a,b] be the domain of f. Then f has a maximum and a minimum.

Proof Let n be a positive integer and consider the finite partition

$$a, \; a + \delta, \; a + 2\delta, \; \ldots, \; a + n\delta = b$$

where $\delta = (b - a)/n$. Let $f(a + m\delta)$ be the greatest of the values

$$f(a), \; f(a + \delta), \; \ldots, \; f(a + n\delta),$$

and put $m = g(n)$. Then every real solution of

$$n \in Z, \; 0 < n, \; \delta = (b - a)/n, \; m = g(n) \tag{4}$$

is a solution of

$$a \leq a + m\delta \leq b. \tag{4'}$$

Furthermore, every real solution of

$$\text{(the system (4))}, \; k \in Z, \; 0 \leq k \leq n \tag{5}$$

is a solution of

$$f(a + mt) \geq f(a + kt). \tag{5'}$$

Let n_1 be a positive infinite hyperinteger and let $\delta_1 = (b - a)/n_1$, $m_1 = g(n_1)$. We show that f has a maximum at

$$c = st(a + m_1\delta_1).$$

The triple (n_1, δ_1, m_1) is a hyperreal solution of (4) and hence of (4'), so

$$a \le a + m_1\delta_1 \le b.$$

Taking standard parts,

$$a \le c \le b.$$

Consider any real number $x \in [a,b]$. By Corollary 2 in Section 3B, x belongs to an infinitesimal subinterval of [a,b] of the form

$$[a + k_1\delta_1, a + (k_1 + 1)\delta_1)^*$$

where k_1 is a hyperinteger between 0 and n_1. Then

$$x = st(a + k_1\delta_1).$$

The quadruple $(n_1, \delta_1, m_1, k_1)$ is a hyperreal solution of (5) and hence of (5'). Thus

$$f(a + m_1\delta_1) \ge f(a + k_1\delta_1).$$

Since f is continuous on [a,b],

$$f(c) = st(f(a + m_1\delta_1)) \ge st(f(a + k_1\delta_1)) = f(x).$$

Thus f has a maximum at c. ⊣

CRITICAL POINT THEOREM

Suppose the domain of f is an interval I, f is continuous on I, and f has a maximum or minimum at a point c in I. Then one of the following occurs:

(i) c is an endpoint of I,
(ii) f'(c) is undefined,
(iii) f'(c) = 0.

Proof Assume neither (i) nor (ii) holds, so f is not an endpoint of I and f'(c) exists. We show (iii). Suppose f has a maximum at c. Let

$\Delta x > 0$ be infinitesimal. Then

$$f(c + \Delta x) \leq f(c), \; f(c - \Delta x) \leq f(c),$$

$$\frac{f(c + \Delta x) - f(c)}{\Delta x} \leq 0 \leq \frac{f(c - \Delta x) - f(c)}{\Delta x},$$

$$st\left(\frac{f(c + \Delta x) - f(c)}{\Delta x}\right) \leq 0 \leq st\left(\frac{f(c - \Delta x) - f(c)}{\Delta x}\right),$$

$$f'(c) \leq 0 \leq f'(c),$$

$$f'(c) = 0.$$ ⊣

We call a point c where (i), (ii), or (iii) occurs a <u>critical point</u> of f. A critical point which is not an endpoint of I is called an <u>interior critical point</u> of f.

MEAN VALUE THEOREM

Assume f is continuous on the closed interval [a,b] and differentiable on (a,b). Then there is a point $c \in (a,b)$ such that the slope at c equals the average slope,

$$f'(c) = \frac{f(b) - f(a)}{b - a}.$$

The proof of the Mean Value Theorem from the Extreme Value and Critical Point Theorems is easy and uses standard methods only, so we shall not give it here. The Intermediate, Extreme, and Mean Value Theorems have the following useful consequences which involve hyperreal numbers.

Suppose f is a real function continuous on an interval I, and $x < y$ in I^*.

HYPERREAL INTERMEDIATE VALUE THEOREM

If u is a hyperreal number between f(x) and f(y), then there is a hyperreal z such that $x \leq z \leq y$ and $f(z) = u$.

HYPERREAL EXTREME VALUE THEOREM

f has a maximum and minimum on the hyperreal closed interval $[x,y]^*$. That is, there is a z between x and y such that whenever $x \le u \le y$, $f(z) \ge f(u)$.

HYPERREAL MEAN VALUE THEOREM

Let f be differentiable on the interior of I. There is a z such that

$$x < z < y \text{ and } f'(z) = \frac{f(y) - f(x)}{y - x}.$$

Here $f'(z)$ means $(f')^*(z)$, the natural extension of f' evaluated at z.

Proof To illustrate the method we prove the Hyperreal Extreme Value Theorem. By the real Extreme Value Theorem, on every closed real subinterval $[x_0,y_0]$ of I, f has a maximum at some point $z_0 = g(x_0,y_0)$. Thus each real solution of

$$x_0 \in I,\ y_0 \in I,\ x_0 \le u_0 \le y_0 \tag{6}$$

is a solution of

$$x_0 \le g(x_0,y_0) \le y_0, f(g(x_0,y_0)) \ge f(u_0). \tag{7}$$

Since every hyperreal solution of (6) is also a solution of (7), f has a maximum in $[x,y]^*$ at $z = g(x,y)$. ⊣

We conclude this section with two applications of the Hyperreal Mean Value Theorem.

SECOND DERIVATIVE TEST

(i) If $f'(c) = 0$ and $f''(c) < 0$, then f has a local maximum at c.
(ii) If $f'(c) = 0$ and $f''(c) > 0$, then f has a local minimum at c.

Proof We prove (i). Since $f''(c)$ exists, f and f' exist on some neighborhood of c. Let $x \approx c$. By Theorem 2 it suffices to prove that $f(c) \ge f(x)$. We assume $f(c) < f(x)$ and obtain a contradiction. Say $c < x$. By the Hyperreal Mean Value Theorem there is a point t such that

$$c < t < x, \; f'(t) = \frac{f(x) - f(c)}{x - c} .$$

Then $f'(t) > 0$. Since $f'(c) = 0$, we have

$$\frac{f'(t) - f'(c)}{t - c} = \frac{f'(t)}{t - c} > 0 .$$

Taking standard parts,

$$f''(c) = st\left(\frac{f'(t) - f'(c)}{t - c}\right) \geq 0 .$$

This contradicts the hypothesis $f''(c) < 0$. The case $c > x$ is similar. We conclude that $f(c) \geq f(x)$. ⊣

The following theorem shows that f has a continuous derivative if and only if the hyperreal formula for $f'(x)$ is valid for hyperreal as well as for real values of x.

THEOREM 3

Let f be a real function which is differentiable on an open interval I. The following are equivalent.

(i) f is continuously differentiable on I. That is, whenever $c \in I$ and $x \approx c$, $f'(x) \approx f'(c)$.

(ii) Whenever $c \in I$, $x \approx c$, and Δx is nonzero infinitesimal,

$$f'(c) \approx \frac{f(x + \Delta x) - f(x)}{\Delta x} .$$

Proof We first show that (i) implies (ii). Assume (i). Let Δx be nonzero infinitesimal. By the Hyperreal Mean Value Theorem there is a hyperreal number t between x and $x + \Delta x$ such that

$$f'(t) = \frac{f(x + \Delta x) - f(x)}{\Delta x} .$$

Since $t \approx x \approx c$, we have $f'(c) \approx f'(t)$ and (ii) follows.

Assume (ii). Every real solution of

$$x \in I, \; \varepsilon > 0 \tag{8}$$

is a partial hyperreal solution of

$$0 < \Delta x < \varepsilon, \quad \left| f'(x) - \frac{f(x + \Delta x) - f(x)}{\Delta x} \right| < \varepsilon, \tag{9}$$

for we may take Δx infinitesimal. Let $x_1 \approx c \in I$ and let ε_1 be positive infinitesimal. Since I is open, $x_1 \in I^*$, hence x_1, ε_1 is a hyperreal solution of (8). By the Partial Solution Theorem there is a Δx_1 such that (9) holds. Then

$$f'(x_1) \approx \frac{f(x_1 + \Delta x_1) - f(x_1)}{\Delta x_1}.$$

But $\Delta x_1 \approx 0$, so by (ii),

$$f'(c) \approx \frac{f(x_1 + \Delta x_1) - f(x_1)}{\Delta x_1}.$$

Therefore $f'(x_1) \approx f'(c)$, and (i) holds. ⊣

The following corollary for closed intervals is sometimes useful.

COROLLARY 4 Suppose the derivative of f is continuous on [a,b]. Then for any infinitesimal subinterval $[x, x + \Delta x]^* \subseteq [a,b]^*$,

$$f'(x) \approx \frac{f(x + \Delta x) - f(x)}{\Delta x}.$$

<u>Proof</u> By Theorem 7, Section 3A, there is a function g which has a continuous derivative everywhere and agrees with f on [a,b]. The corollary follows by applying Theorem 3 to g. ⊣

INTEGRATION 4

In this chapter we use hyperreal numbers to develop the Riemann integral. To keep the theory as elementary as possible, we restrict ourselves to continuous real functions.

PERMANENT ASSUMPTION

We assume throughout this chapter that f and g are real functions which are continuous on an interval I.

4A THE DEFINITE INTEGRAL (§4.1 - §4.3)

Given a positive real function f, consider the region bounded by the lines $x = a$, $x = b$, $y = 0$, and the curve $y = f(x)$. We call this region the region under $y = f(x)$ from a to b. The area of this region may be regarded as a function $A(a,b)$ of two variables. In this and the next section we shall define the definite integral

$$\int_a^b f(x)\,dx$$

and show that it is equal to the area under $y = f(x)$ from a to b. Our plan is as follows. First we list some properties which the intuitive concept of area has. Second we prove that the definite integral has these properties. Third we prove that the definite integral is the only function with these properties.

DEFINITION

By an <u>area function</u> for f we mean a real function A(u,w), whose domain is the set of ordered pairs of elements of I, such that

(i) A has the Addition Property:

$$A(a,c) = A(a,b) + A(b,c) \text{ for all } a,b,c \in I.$$

(ii) A has the Rectangle Property:

$$m(b - a) \leq A(a,b) \leq M(b - a)$$

whenever $a < b$ in I and f has minimum value m and maximum value M on [a,b].

The Rectangle Property states that the area of the region is between the areas of the inscribed and circumscribed rectangles. It follows at once from the Addition Property that

$$A(a,a) = 0, \; A(b,a) = -A(a,b).$$

Thus to specify an area function we need only specify the values of A(a,b) for $a < b$ in I.

We now introduce the notion of a Riemann sum. For simplicity we shall consider only partitions of [a,b] in which all subintervals except the last subinterval have the same length.

DEFINITION

Let [a,b] be a subinterval of I and let Δx be a positive real number. The <u>Riemann sum</u> $\sum_a^b f(x)\Delta x$ is defined as the sum

$$\sum_a^b f(x)\Delta x = f(x_0)\Delta x + f(x_1)\Delta x + \dots + f(x_{n-1})\Delta x + f(x_n)(b - x_n)$$

where n is the largest integer such that $a + n\Delta x \leq b$, and

$$x_0 = a, \; x_1 = a + \Delta x, \; \dots, \; x_n = a + n\Delta x.$$

Geometrically, the interval [a,b] is partitioned into subintervals of equal length Δx, except that if Δx does not evenly divide $b - a$ then the last subinterval $[x_n, b]$ is shorter than Δx. The Riemann sum is equal to the sum of the areas of the vertical strips over each subinterval with height equal to the value of $f(x)$ at the left end of the subinterval.

The Riemann sum $\sum_a^b f(x)\Delta x$ is a real function of the three variables a, b, Δx. If a and b are held fixed it becomes a function of the single variable Δx. If we replace the positive real Δx by a positive infinitesimal dx, the natural extension gives us the <u>infinite Riemann sum</u>

$$\sum_a^b f(x)dx.$$

Since the finite Riemann sum is defined for all real $\Delta x > 0$, the infinite Riemann sum is defined for all hyperreal $dx > 0$. Our plan is to define the integral as the standard part of the infinite Riemann sum. First we must prove that this sum is finite, so its standard part exists.

THEOREM 1

Let $a < b$ in I and let dx be positive infinitesimal. Then the infinite Riemann sum $\sum_a^b f(x)dx$ is a finite hyperreal number.

<u>Proof</u> Let m be the minimum and M the maximum value of f on [a,b]. For each positive real Δx we have

$$\sum_a^b m\Delta x \le \sum_a^b f(x)\Delta x \le \sum_a^b M\Delta x ,$$

and

$$\sum_a^b m\Delta x = m(b - a), \ \sum_a^b M\Delta x = M(b - a).$$

Therefore every real solution of $\Delta x > 0$ is a solution of

$$m(b - a) \le \sum_a^b f(x)\Delta x \le M(b - a). \qquad (1)$$

Since $dx > 0$, dx is a hyperreal solution of (1), and therefore $\sum_a^b f(x)dx$ is finite. ⊣

DEFINITION

Let $a < b$ in I and let dx be a positive infinitesimal. The definite integral of f from a to b with respect to dx is standard part of the Riemann sum,

$$\int_a^b f(x)dx = st(\sum_a^b f(x)dx) \ .$$

Moreover,

$$\int_a^a f(x)dx = 0, \quad \int_b^a f(x)dx = -\int_a^b f(x)dx.$$

For each fixed positive infinitesimal dx, the definite integral

$$\int_u^w f(x)dx$$

is a real function of two variables u and w. It does not depend on the dummy variable x. We always use matching symbols for the dummy variable x and the infinitesimal dx. This convention identifies the dummy variable when integrating a function of two or more variables. Thus

$$\int_0^1 x^2 \ t \ dx = \frac{1}{3}t, \quad \int_0^1 x^2 \ t \ dt = \frac{1}{2}x^2.$$

We now develop some properties of the definite integral.

THEOREM 2

Let $a < b$ in I, let c be a real constant, and let dx be positive infinitesimal. Then

(i) $\int_a^b c \ dx = c(b - a)$

(ii) $\int_a^b c \ f(x)dx = c\int_a^b f(x)dx$

(iii) $\int_a^b (f(x) + g(x))dx = \int_a^b f(x)dx + \int_a^b g(x)dx$

(iv) If $f(x) \le g(x)$ for all $x \in [a,b]$, $\int_a^b f(x)dx \le \int_a^b g(x)dx$.

Proof In each case the proof has three steps. First, verify the analogous formula for finite Riemann sums. Second, use the Solution Axiom to prove the formula for infinite Riemann sums. Third, take standard parts. ⊣

The following theorem shows that the definite integral $\int_a^b f(x)dx$ does not depend on the infinitesimal dx.

THEOREM 3

Let $a < b$ in I and let dx and du be two positive infinitesimals. Then

$$\int_a^b f(x)dx = \int_a^b f(u)du.$$

Hence

$$\int_a^b f(x)dx = \lim_{\Delta x \to 0^+} \sum_a^b f(x)\Delta x \ .$$

Proof It suffices to prove that for every positive real number r,

$$\int_a^b f(x)dx \le \int_a^b f(u)du + r \ .$$

Let $c = r/(b - a)$. We shall show that

$$\sum_a^b f(x)dx \le \sum_a^b (f(u) + c)du \ , \tag{2}$$

whence by Theorem 2,

$$\int_a^b f(x)dx \le \int_a^b (f(u) + c)du = \int_a^b f(u)du + r \ .$$

In Elementary Calculus, Formula (2) was justified intuitively. To give a rigorous proof of (2) we use the Partial Solution Theorem. Let Δx and Δu be positive real numbers. If

$$\sum_a^b f(x)\Delta x > \sum_a^b (f(u) + c)\Delta u ,$$

there must be a point at which a rectangle in $\sum_a^b f(x)\Delta x$ is above a rectangle in $\sum_a^b (f(u) + c)\Delta u$. So there must be a pair of points x, u in [a,b] such that

$$x - \Delta u \le u \le x + \Delta x, \; f(x) > f(u) + c.$$

Thus every real solution of

$$\Delta x > 0, \; \Delta u > 0, \; \sum_a^b f(x)\Delta x > \sum_a^b (f(u) + c)\Delta u \tag{3}$$

is a partial real solution of

$$a \le x \le b, \; a \le u \le b, \; x - \Delta u \le u \le x + \Delta x, \; f(x) > f(u) + c . \tag{4}$$

Now suppose (2) fails for dx and du, so $\sum_a^b f(x)dx > \sum_a^b (f(u) + c)du$. We have a hyperreal solution of (3) and hence of (4) for some x_1, u_1. But since dx and du are infinitesimal, (4) implies that $x_1 \approx u_1$ and $f(x_1) \not\approx f(u_1)$, contradicting the continuity of f. We conclude that (2) is true. ⊣

From now on when we write $\int_a^b f(x)dx$, it is understood that dx is some positive infinitesimal. By Theorem 3, it doesn't matter which one.

THEOREM 4

The definite integral $\int_a^b f(x)dx$ is an area function for f.

<u>Proof</u> The Rectangle Property follows at once from Theorem 2.

$$m \le f(x) \le M,$$

$$m(b - a) = \int_a^b m dx \le \int_a^b f(x)dx \le \int_a^b M dx = M(b - a).$$

By Theorem 3, it suffices to prove the Addition Property for one positive infinitesimal dx. Let a < b < c in I. Let n be a positive integer and let $\Delta x = (b - a)/n$. Then $b = n\Delta x$ is a point in the partition of [a,c], so

$$\sum_a^c f(x)\Delta x = \sum_a^b f(x)\Delta x + \sum_b^c f(x)\Delta x \ . \qquad (5)$$

Thus every real solution of

$$n \in Z,\ 0 < n,\ \Delta x = (b - a)/n \qquad (6)$$

is a solution of (5). Now let n_1 be a positive infinite hyperinteger and let $dx = (b - a)/n_1$. We have a hyperreal solution of (6) and hence of (5), so

$$\sum_a^c f(x)dx = \sum_a^b f(x)dx + \sum_b^c f(x)dx \ .$$

Since dx is positive infinitesimal, we may take standard parts and obtain the Addition Property for dx,

$$\int_a^c f(x)dx = \int_a^b f(x)dx + \int_b^c f(x)dx \ .$$

⊣

4B FUNDAMENTAL THEOREM OF CALCULUS (§4.4, §4.5)

We shall prove that the definite integral is the unique area function for f, and use that fact to prove the Fundamental Theorem of Calculus. We first introduce lower and upper Riemann sums. For simplicity we consider only the case where Δx evenly divides the interval [a,b].

DEFINITION

Let $\Delta x = (b - a)/n$ for some positive integer n. The <u>lower Riemann sum</u> of f is the sum

$$\sum_a^b m(x, x + \Delta x)\Delta x = m(x_0, x_1)\Delta x + m(x_1, x_2)\Delta x + \dots + m(x_{n-1}, x_n)\Delta x \ ,$$

and the <u>upper Riemann sum</u> of f is the sum

$$\sum_a^b M(x,x+\Delta x)\Delta x = M(x_0,x_1)\Delta x + M(x_1,x_2)\Delta x + \dots + M(x_{n-1},x_n)\Delta x,$$

where

$$m(x,x+\Delta x) = \text{minimum value of } f \text{ on } [x,x+\Delta x],$$

$$M(x,x+\Delta x) = \text{maximum value of } f \text{ on } [x,x+\Delta x],$$

$$x_k = a + k\Delta x.$$

Geometrically, the lower Riemann sum is the sum of the inscribed rectangles and the upper Riemann sum is the sum of the circumscribed rectangles. For a given real function f continuous on [a,b], the lower and upper Riemann sums

$$\sum_a^b m(x,x+\Delta x)\Delta x,\ \sum_a^b M(x,x+\Delta x)\Delta x,\ \Delta x = (b-a)/n$$

are real functions of n, and are defined whenever n is a positive integer. By the Solution Axiom, their natural extensions are defined for any positive hyperinteger. Given a positive infinite hyperinteger H, we let $dx = (b-a)/H$ and call the natural extensions

$$\sum_a^b m(x,x+dx)dx,\quad \sum_a^b M(x,x+dx)dx$$

the <u>infinite lower and upper Riemann sums</u> of f with respect to dx.

LEMMA 1 The infinite lower and upper Riemann sums are infinitely close to each other,

$$\sum_a^b m(x,x+dx)dx \approx \sum_a^b M(x,x+dx)dx.$$

<u>Proof</u> For each positive integer n with $\Delta x = (b-a)/n$, we have

$$\sum_a^b m(x,x+\Delta x)\Delta x \le \sum_a^b M(x,x+\Delta x)\Delta x,$$

so by the Solution Axiom,

$$\sum_a^b m(x,x + dx)dx \le \sum_a^b M(x,x + dx)dx.$$

We must show that for each positive real r,

$$\sum_a^b M(x,x + dx)dx < \sum_a^b m(x,x + dx)dx + r. \tag{1}$$

The proof is like that of Theorem 3 in the preceding section. Let $c = r/(b - a)$. Consider a positive integer n with $\Delta x = (b - a)/n$. If

$$\sum_a^b M(x,x + \Delta x)\Delta x \ge \sum_a^b m(x,x + \Delta x)\Delta x + r,$$

then there must be an x such that

$$M(x,x + \Delta x) \ge m(x,x + \Delta x) + c.$$

But $M(x,x + \Delta x) = f(y)$ and $m(x,x + \Delta x) = f(z)$ for some $y,z \in [x,x + \Delta x]$. Thus any real solution of

$$\begin{gathered} n \in Z,\ 0 < n,\ \Delta x = (b - a)/n, \\ \sum_a^b M(x,x + \Delta x)\Delta x \ge \sum_a^b m(x,x + \Delta x)\Delta x + r \end{gathered} \tag{2}$$

is a partial real solution of

$$a \le y \le b,\ a \le z \le b,\ |y - z| \le \Delta x,\ f(y) \ge f(z) + c. \tag{3}$$

Now let $dx = (b - a)/n_1$ where n_1 is a positive infinite hyperinteger. Suppose (1) fails,

$$\sum_a^b M(x,x + dx)dx \ge \sum_a^b m(x,x + dx)dx + r.$$

We have a hyperreal solution of (2) and hence of (3) for some y_1, z_1. It follows that

$$y_1 \approx z_1,\ f(y_1) \not\approx f(z_1),$$

contradicting the continuity of f. We conclude that (1) holds. ⊣

THEOREM 2

There is only one area function for f, so the definite integral is the unique area function.

Proof Let A(u,v) and B(u,v) be area functions for f. Let $a < b$ in I. For each positive integer n with $\Delta x = (b - a)/n$, the Rectangle Property gives

$$m(x_k,x_{k+1})\Delta x \le A(x_k,x_{k+1}) \le M(x_k,x_{k+1})\Delta x,$$

$$k = 0,1,\ldots,n-1,$$

and by the Addition Property,

$$\sum_a^b m(x,x+\Delta x)\Delta x \le A(a,b) \le \sum_a^b M(x,x+\Delta x)\Delta x.$$

Let n_1 be a positive infinite hyperinteger and $dx = (b - a)/n_1$. By the Solution Axiom,

$$\sum_a^b m(x,x+dx)dx \le A(a,b) \le \sum_a^b M(x,x+dx)dx.$$

Similarly,

$$\sum_a^b m(x,x+dx)dx \le B(a,b) \le \sum_a^b M(x,x+dx)dx.$$

By Lemma 1,

$$A(a,b) \approx B(a,b),$$

and since A(a,b) and B(a,b) are real they must be equal. If $b \le a$ they are still equal because $A(a,a) = B(a,a) = 0$, and

$$A(b,a) = -A(a,b) = -B(a,b) = B(b,a).$$ ⊣

DEFINITION

Suppose the domain of f is an open interval I. F is said to be an antiderivative of f on I if F has domain I and f is the derivative of F.

FUNDAMENTAL THEOREM OF CALCULUS

Suppose $a < b$ in I and F is an antiderivative of f. Then

$$\int_a^b f(x)dx = F(b) - F(a).$$

<u>Proof</u> Let $D(a,b) = F(b) - F(a)$. We show that D is an area function for f.

Addition Property:

$$D(a,c) = F(c) - F(a) = (F(b) - F(a)) + (F(c) - F(b))$$
$$= D(a,b) + D(b,c).$$

Rectangle Property: By the Mean Value Theorem there is a point $c \in (a,b)$ such that

$$f(c) = F'(c) = \frac{F(b) - F(a)}{b - a},$$

$$f(c)(b - a) = F(b) - F(a) = D(a,b).$$

Let m and M be the minimum and maximum values of f on [a,b]. Then

$$m \le f(c) \le M,$$

$$m(b - a) \le f(c)(b - a) \le M(b - a),$$

$$m(b - a) \le D(a,b) \le M(b - a). \qquad \dashv$$

THEOREM 3

Suppose the domain of f is an open interval I and F is an antiderivative of f. Then the set of all antiderivatives of F is equal to the set of all functions which differ from F by a constant.

<u>Proof</u> For any constant C_0, $F(x) + C_0$ has the same derivative as F and hence is an antiderivative of f. Suppose G is an antiderivative of f, and let $H(x) = F(x) - G(x)$. We prove that H is constant. We have

$$H'(x) = F'(x) - G'(x) = f(x) - f(x) = 0.$$

By the Mean Value Theorem, whenever $a < b$ in I there is a point c (a,b) such that

$$0 = H'(c) = \frac{H(b) - H(a)}{b - a}.$$

Therefore $H(b) = H(a)$, and H is a constant function. ⊣

DEFINITION

The set of all antiderivatives of f is called the <u>indefinite integral</u> of f, and is written $\int f(x)dx$.

The set of all functions which differ from F by a constant is denoted by

$$F(x) + C.$$

If F is an antiderivative of f, then Theorem 3 states that

$$\int f(x)dx = F(x) + C.$$

The techniques for evaluating integrals are exactly the same as in the standard calculus treatment. As usual, one can evaluate a definite integral by finding an antiderivative and using the Fundamental Theorem of Calculus. Notice that we have not yet proved that every continuous real function f has an antiderivative. This will be done in the next section.

4C SECOND FUNDAMENTAL THEOREM OF CALCULUS (§4.8)

We may regard the definite integral $\int_u^w f(t)dt$ as a real function of two variables defined for all $u, w \in I$. By the Function Axiom, $\int_u^w f(t)dt$ is also defined when u and w are hyperreal numbers in I^*.

SECOND FUNDAMENTAL THEOREM OF CALCULUS

Let $a \in I$ and define $F(x)$ by

$$F(x) = \int_a^x f(t)dt.$$

(i) F is continuous on I.

(ii) F is an antiderivative of f on the interior of I.

<u>Proof</u> (i) Let $c \in I$ and let Δx be an infinitesimal such that $c + \Delta x \in I^*$. We must show that

$$F(c + \Delta x) \approx F(c). \tag{1}$$

Suppose $\Delta x > 0$, and let b be a point in I such that $c < c + \Delta x < b$. Let m be the minimum and M the maximum value of f in $[c,b]$. For each positive real number $u < b - c$ we have

$$F(c + u) - F(c) = \int_c^{c+u} f(t)dt,$$

$$mu \le \int_c^{c+u} f(t)dt \le Mu$$

and hence

$$mu \le F(c + u) - F(c) \le Mu.$$

By the Solution Axiom we have

$$m\Delta x \le F(c + \Delta x) - F(c) \le M\Delta x,$$

and (1) follows. The case $\Delta x < 0$ is similar.

(ii) Let c be an interior point of I and let Δx be nonzero infinitesimal. Again suppose $\Delta x > 0$ and let $b \in I$, $c < c + \Delta x < b$. Consider a real number $u \in (0,b - c)$. On the interval $[c,c + u]$ let f have a minimum at y and a maximum at z. Then

$$F(c + u) - F(c) = \int_c^{c+u} f(t)dt,$$

$$f(y)u \le \int_c^{c+u} f(t)dt \le f(z)u.$$

Thus any real solution of

$$0 < u < b - c \tag{2}$$

is a partial real solution of

$$\left.\begin{aligned} & c \le y \le c + u,\ c \le z \le c + u, \\ & f(y)u \le F(c + u) - F(c) \le f(z)u. \end{aligned}\right\} \tag{3}$$

$u = \Delta x$ is a hyperreal solution of (2) and hence a partial hyperreal solution of (3). Thus there are hyperreal numbers y_1, z_1 such that

$$y_1 \approx c,\ z_1 \approx c,\ f(y_1)\Delta x \le F(c + \Delta x) - F(c) \le f(z_1)\Delta x.$$

Then

$$f(y_1) \le \frac{F(c + \Delta x) - F(c)}{\Delta x} \le f(z_1) .$$

Using the continuity of f,

$$\begin{aligned} f(c) = st(f(y_1)) &\le st\left(\frac{F(c + \Delta x) - F(c)}{\Delta x}\right) \\ &\le st(f(z_1)) = f(c), \end{aligned}$$

whence

$$f(c) = st\left(\frac{F(c + \Delta x) - F(c)}{\Delta x}\right).$$

A similar argument works when $\Delta x < 0$. It follows that

$$f(c) = F'(c).$$

⊣

The Second Fundamental Theorem gives us another proof of the Fundamental Theorem of Calculus in Section 4B. This proof does not depend on the fact that the area function for f is unique.

Alternate Proof of the Fundamental Theorem of Calculus Let G be an antiderivative of f. Since any two antiderivatives differ by a constant,

$$G(x) = \int_a^x f(t)dt + C_0$$

for some constant C_0. Then

$$\begin{aligned} G(b) - G(a) &= \left(\int_a^b f(t)dt + C_0\right) - \left(\int_a^a f(t)dt + C_0\right) \\ &= \int_a^b f(t)dt. \end{aligned}$$

⊣

LIMITS 5

5A EQUIVALENCE THEOREMS (§5.1, §5.4)

We show that the infinitesimal definitions of limit, continuity, and uniform continuity are equivalent to the standard ε,δ definitions. These equivalence theorems will be useful later on.

THEOREM 1

Let f be a real function and c a real number. The following are equivalent.

(i) $\lim_{x \to c} f(x) = L$. That is, whenever $x \approx c$ but $x \neq c$, $f(x) \approx L$.

(ii) There exists a hyperreal $\delta > 0$ such that whenever $0 < |x - c| < \delta$, $f(x) \approx L$.

(iii) The ε,δ condition: For every real $\varepsilon > 0$ there is a real $\delta > 0$ such that whenever x is real and $0 < |x - c| < \delta$, $|f(x) - L| < \varepsilon$.

<u>Proof</u> (i) obviously implies (ii), with δ being any positive infinitesimal. To prove (ii) implies (iii), assume (iii) fails for some real $\varepsilon > 0$. Then every real $\delta > 0$ is a partial real solution of

$$0 < |x - c| < \delta, \quad |f(x) - L| \geq \varepsilon. \tag{1}$$

Let $\delta_1 > 0$ be hyperreal. By the Partial Solution Theorem, there is a hyperreal x_1 such that (1) holds, and therefore

$$0 < |x_1 - c| < \delta_1, \quad f(x_1) \not\approx L.$$

Thus if (iii) fails then (ii) fails, so (ii) implies (iii).

Assume (iii). Let $x_1 \approx c$ but $x_1 \neq c$. Let ε be any positive real number, and let δ be the corresponding positive real number in the ε,δ condition. Then every real solution of $0 < |x - c| < \delta$ is a solution of $|f(x) - L| < \varepsilon$. We have $0 < |x_1 - c| < \delta$, and therefore $|f(x_1) - L| < \varepsilon$. Since this holds for all positive real ε, $f(x_1) \approx L$. ⊣

Condition (ii) in the above theorem is a priori weaker than (i) and sometimes easier to verify.

COROLLARY 2 The following are equivalent.

(i) f is continuous at c. That is, whenever $x \approx c$, $f(x) \approx f(c)$.

(ii) There is a hyperreal $\delta > 0$ such that whenever $|x - c| < \delta$, $f(x) \approx f(c)$.

(iii) The ε,δ condition: For every real $\varepsilon > 0$ there is a real $\delta > 0$ such that whenever x is real and $|x - c| < \delta$, $|f(x) - f(c)| < \varepsilon$.

COROLLARY 3 The following are equivalent.

(i) $f'(c) = S$. That is, whenever $x \approx c$ but $x \neq c$,

$$\frac{f(c + \Delta x) - f(c)}{\Delta x} \approx S.$$

(ii) There is a hyperreal $\delta > 0$ such that whenever $0 < |\Delta x| < \delta$,

$$\frac{f(c + \Delta x) - f(c)}{\Delta x} \approx S.$$

(iii) The ε,δ condition: For every real $\varepsilon > 0$ there is a real $\delta > 0$ such that whenever Δx is real and $0 < |\Delta x| < \delta$,

$$\left|\frac{f(c + \Delta x) - f(c)}{\Delta x} - S\right| < \varepsilon.$$

We shall now state without proof the forms of Theorem 1 and Corollary 2 restricted to a set $Y \subsetneq R$.

THEOREM 1'

Let $Y \subseteq R$ and $c \in Y$. The following are equivalent.

(i) $\lim_{\substack{x \to c \\ x \in Y}} f(x) = L$. That is, whenever $x \in Y^*$, $x \approx c$, $x \neq c$, we have $f(x) \approx L$.

(ii) There is a hyperreal $\delta > 0$ such that whenever $x \in Y^*$ and $0 < |x - c| < \delta$, $f(x) \approx L$.

(iii) The ε, δ condition: For every real $\varepsilon > 0$ there is a real $\delta > 0$ such that whenever $x \in Y$ and $0 < |x - c| < \delta$, $|f(x) - L| < \varepsilon$.

COROLLARY 2' The following are equivalent.

(i) f is continuous on Y. That is, whenever $c \in Y$, $x \in Y^*$, $x \approx c$, $x \neq c$, we have $f(x) \approx f(c)$.

(ii) For each $c \in Y$ there exists a hyperreal $\delta > 0$ such that whenever $x \in Y^*$ and $0 < |x - c| < \delta$, $f(x) \approx f(c)$.

(iii) The ε, δ condition: For every real $\varepsilon > 0$ and $c \in Y$ there is a real $\delta > 0$ such that whenever $x \in Y$ and $0 < |x - c| < \delta$, $|f(x) - f(c)| < \varepsilon$.

We next prove an equivalence theorem for uniform continuity.

THEOREM 4

The following are equivalent.

(i) f is uniformly continuous on Y. That is, whenever $x, y \in Y^*$ and $x \approx y$, $f(x) \approx f(y)$.

(ii) There is a hyperreal $\delta > 0$ such that whenever $x, y \in Y^*$ and $|x - y| < \delta$, $f(x) \approx f(y)$.

(iii) The ε, δ condition: For every real $\varepsilon > 0$ there is a real $\delta > 0$ such that whenever $x, y \in Y$ and $|x - y| < \delta$, $|f(x) - f(y)| < \varepsilon$.

<u>Proof</u> (i) trivially implies (ii). Assume (iii) fails for some real $\varepsilon > 0$. Then every real $\delta > 0$ is a partial real solution of

$$x \in Y,\ y \in Y,\ |x - y| < \delta,\ |f(x) - f(y)| \geq \varepsilon. \tag{2}$$

Hence every hyperreal $\delta_1 > 0$ is a partial hyperreal solution of (2).

So there exist $x_1, y_1 \in Y^*$ such that $|x_1 - y_1| < \delta_1$ but $f(x_1) \not\approx f(y_1)$. This shows that the failure of (iii) implies the failure of (ii), so (ii) implies (iii).

Assume (iii). Let $x_1, y_1 \in Y^*$ and $x_1 \approx y_1$. Let ε be any positive real and let $\delta > 0$ be the corresponding number in the ε,δ condition. Every real solution of

$$x \in Y,\ y \in Y,\ |x - y| < \delta \tag{3}$$

is a solution of

$$|f(x) - f(y)| < \varepsilon. \tag{4}$$

(x_1, y_1) is a hyperreal solution of (3) and hence of (4). Since this holds for all real $\varepsilon > 0$, $f(x_1) \approx f(y_1)$. ⊣

We conclude with a discussion of infinite limits.

DEFINITION

$\lim_{x\to\infty} f(x) = L$ if $f(H) \approx L$ for every positive infinite H.

$\lim_{x\to c} f(x) = \infty$ if $f(x)$ is positive infinite whenever $x \approx c$ but $x \neq c$.

We state equivalence theorems for these limits without proof.

THEOREM 5

The following are equivalent.

(i) $\lim_{x\to\infty} f(x) = L$.

(ii) There is a hyperreal number K such that whenever $H > K$, $f(H) \approx L$.

(iii) The ε,M condition: For every real $\varepsilon > 0$ there is a real number M such that whenever x is real and $x > M$, $|f(x) - L| < \varepsilon$.

THEOREM 6

The following are equivalent.

(i) $\lim_{x \to c} f(x) = \infty$.

(ii) There is a hyperreal $\delta > 0$ such that whenever $0 < |x - c| < \delta$, $f(x)$ is positive infinite.

(iii) The M, δ condition: For every real number M there is a real $\delta > 0$ such that whenever x is real and $0 < |x - c| < \delta$, $f(x) > M$.

Negative infinite limits, infinite limits restricted to a set Y, and limits such as $\lim_{x \to \infty} f(x) = \infty$, are defined analogously and have similar equivalence theorems.

5B L'HOSPITAL'S RULE (§5.2)

The proof of l'Hospital's Rule uses the Generalized Mean Value Theorem.

GENERALIZED MEAN VALUE THEOREM

Suppose f and g are continuous on the closed interval [a,b] and differentiable on the open interval (a,b). Assume further that $g'(x) \neq 0$ for $x \in (a,b)$. Then there is a point $t \in (a,b)$ such that

$$\frac{f'(t)}{g'(t)} = \frac{f(b) - f(a)}{g(b) - g(a)}.$$

L'HOSPITAL'S RULE FOR 0/0

Suppose that for all x in some real open interval (c,b), $f'(x)$ and $g'(x)$ exist and $g'(x) \neq 0$. Assume that

$$\lim_{x \to c^+} f(x) = 0, \quad \lim_{x \to c^+} g(x) = 0.$$

If $\lim_{x \to c^+} \frac{f'(x)}{g'(x)}$ exists then

$$\lim_{x \to c^+} \frac{f(x)}{g(x)} = \lim_{x \to c^+} \frac{f'(x)}{g'(x)}.$$

<u>Proof</u> Let $\lim_{x \to c^+} \frac{f'(x)}{g'(x)} = L$. We may set $f(c) = 0$ and $g(c) = 0$, so f and g are

continuous on $[c,b)$. By the Generalized Mean Value Theorem, every real solution of

$$c < x < b \tag{1}$$

is a partial real solution of

$$c < t < x, \ \frac{f'(t)}{g'(t)} = \frac{f(x) - f(c)}{g(x) - g(c)},$$

which simplifies to

$$c < t < x, \ \frac{f'(t)}{g'(t)} = \frac{f(x)}{g(x)}. \tag{2}$$

Now let $x_1 \approx c$ and $x_1 > c$. By the Partial Solution Theorem there exists t_1 such that (2) holds. Thus

$$t_1 \neq c, \ t_1 \approx c, \ \frac{f'(t_1)}{g'(t_1)} = \frac{f(x_1)}{g(x_1)}.$$

Since $L = \lim_{t \to c^+} (f'(t)/g'(t))$, we have

$$L \approx \frac{f'(t_1)}{g'(t_1)} = \frac{f(x_1)}{g(x_1)}, \quad L = \lim_{x \to c^+} \frac{f(x)}{g(x)}. \qquad \dashv$$

L'Hospital's Rule also holds when either c or L or both are replaced by ∞ or $-\infty$, and for $x \to c^-$ and $x \to c$, with only routine changes in the proof. The proof of l'Hospital's Rule for ∞/∞ is more difficult.

L'HOSPITAL'S RULE FOR ∞/∞.

Suppose that for all x in some real open interval (c,b), $f'(x)$ and $g'(x)$ exist and $g'(x) \neq 0$. Assume that

$$\lim_{x \to c^+} f(x) = \infty, \ \lim_{x \to c^+} g(x) = \infty.$$

If $\lim_{x \to c^+} \frac{f'(x)}{g'(x)}$ exists then

$$\lim_{x \to c^+} \frac{f(x)}{g(x)} = \lim_{x \to c^+} \frac{f'(x)}{g'(x)}.$$

Proof We shall use the equivalence theorem for limits. Let $L = \lim_{x \to c^+} (f'(x)/g'(x))$. By the Generalized Mean Value Theorem every real solution of

$$c < x < y < c + r \tag{3}$$

is a partial real solution of

$$x < t < y, \; \frac{f'(t)}{g'(t)} = \frac{f(y) - f(x)}{g(y) - g(x)},$$

which we rewrite as

$$x < t < y, \; \frac{f'(t)}{g'(t)} = \frac{\frac{f(x)}{g(x)} - \frac{f(y)}{g(x)}}{1 - \frac{g(y)}{g(x)}}. \tag{4}$$

Let $y_1 \approx c$ and $y_1 > c$. Then $f(y_1)$ and $g(y_1)$ are positive infinite, and so is their product $K = f(y_1)g(y_1)$. By the M, δ condition for $\lim_{x \to c^+} g(x) = \infty$, for every real M there is a real $\delta(M)$ such that every real solution of

$$c < x < c + \delta(M)$$

is a solution of $g(x) > M$. Let δ_1 be such that $0 < \delta_1 \le \delta(K)$ and $c + \delta_1 < y_1$. Consider any x_1 with $c < x_1 < c + \delta_1$. By the Solution Axiom, $g(x_1) > K$. Moreover,

$$c < x_1 < y_1 < c + r,$$

so by the Partial Solution Theorem there is a t_1 such that (4) holds. Then

$$t_1 \approx c, \; \frac{f'(t_1)}{g'(t_1)} \approx L.$$

Also,

$$\left|\frac{f(y_1)}{g(x_1)}\right| \le \left|\frac{f(y_1)}{K}\right| = \left|\frac{f(y_1)}{f(y_1)g(y_1)}\right| \approx 0$$

so $f(y_1)/g(x_1) \approx 0$. Similarly $g(y_1)/g(x_1) \approx 0$. Taking standard parts in (4), we have

$$\mathrm{st}\left(\frac{f'(t_1)}{g'(t_1)}\right) = \mathrm{st}\left(\frac{\dfrac{f(x_1)}{g(x_1)} - \dfrac{f(y_1)}{g(x_1)}}{1 - \dfrac{g(y_1)}{g(x_1)}}\right) = \frac{\mathrm{st}\left(\dfrac{f(x_1)}{g(x_1)}\right) - 0}{1 - 0},$$

whence

$$L \approx \frac{f'(t_1)}{g'(t_1)} \approx \frac{f(x_1)}{g(x_1)}.$$

Since this holds for all $c < x_1 < c + \delta_1$, we see from the Equivalence Theorem for Limits that

$$L = \lim_{x \to c^+} \frac{f(x)}{g(x)}. \qquad \dashv$$

APPLICATIONS OF THE INTEGRAL 6

6A INFINITE SUM THEOREM (§6.1, §6.2, §6.6)

The Infinite Sum Theorem is a simple and extremely useful criterion for a quantity to be equal to the definite integral of a function. It can be used to justify many familiar applications of the integral in geometry and physics. It captures the intuitive idea of obtaining an integration formula by considering a typical infinitesimal element and adding up.

Recall from Chapter 2 that given a nonzero infinitesimal Δx,

$$u \approx v \text{ (compared to } \Delta x)$$

means that

$$\frac{u}{\Delta x} \approx \frac{v}{\Delta x}.$$

INFINITE SUM THEOREM

Assume that

(i) h is a real function continuous on the interval $[a,b]$.

(ii) $B(u,w)$ is a real function which has the Addition Property,

$$B(u,w) = B(u,v) + B(v,w) \text{ for } u < v < w \text{ in } [a,b].$$

(iii) For any infinitesimal subinterval $[x, x + \Delta x]^*$ of $[a,b]^*$,

$$B(x,x+\Delta x) \approx h(x)\Delta x \text{ (compared to } \Delta x).$$

Then

$$B(a,b) = \int_a^b h(x)dx.$$

We shall write

$$\Delta B = B(x,x+\Delta x).$$

The theorem intuitively says that if each infinitesimal piece ΔB is infinitely close to $h(x)\Delta x$ compared to Δx, then the sum $B(a,b)$ of all the ΔB's is infinitely close to the Riemann sum $\Sigma_a^b h(x)\Delta x$.

Proof Let n_1 be a positive infinite hyperinteger and let $\Delta x_1 = (b-a)/n_1$. We shall prove that for every positive real number r,

$$\sum_a^b (h(x) - r)\Delta x_1 < B(a,b) < \sum_a^b (h(x) + r)\Delta x_1.$$

This will show that

$$\sum_a^b h(x)\Delta x_1 - r(b-a) < B(a,b) < \sum_a^b h(x)\Delta x_1 + r(b-a),$$

$$\sum_a^b h(x)\Delta x_1 \approx B(a,b),$$

$$\int_a^b h(x)dx = B(a,b).$$

Consider a positive integer n and let $\Delta x = (b-a)/n$. By the Addition Property,

$$B(a,b) = \sum_a^b B(x,x+\Delta x).$$

It follows that every real solution of

$$n \in Z,\ 0 < n,\ \Delta x = (b-a)/n,\ B(a,b) \geq \sum_a^b (h(x)+r)\Delta x \quad (1)$$

is a partial real solution of

$$a \le x < x + \Delta x \le b,\ B(x,x+\Delta x) \ge (h(x) + r)\Delta x. \tag{2}$$

We may rewrite (2) as

$$a \le x < x + \Delta x \le b,\ \frac{B(x,x+\Delta x)}{\Delta x} \ge h(x) + r. \tag{3}$$

By hypothesis (iii), there is no hyperreal number x_1 such that (3) holds for x_1, Δx_1. By the Partial Solution Theorem, (1) cannot hold for n_1, Δx_1, so

$$B(a,b) < \sum_a^b (h(x) + r)\Delta x_1.$$

The proof that

$$\sum_a^b (h(x) - r)\Delta x_1 < B(a,b)$$

is similar. ⊣

The Infinite Sum Theorem is actually a sharp form of Theorem 2 in Section 4B which states that the definite integral is the unique area function for f. As a first indication of how the Infinite Sum Theorem is used, we give a second proof of the uniqueness of the area function.

THEOREM 2 OF SECTION 4B (Repeated)

A real function f continuous on [a,b] has at most one area function.

Second Proof We show that for any area function $B(u,v)$ for f, $B(a,b) = \int_a^b f(x)dx$. $B(u,v)$ has the Addition Property. Consider any infinitesimal subinterval $[x,x + \Delta x]^*$ of $[a,b]^*$. By the Hyperreal Extreme Value Theorem, f has a minimum value m and a maximum value M in $[x,x + \Delta x]^*$. By the Rectangle Property and Solution Axiom,

$$m\Delta x \le \Delta B \le M\Delta x,$$

$$m \le \Delta B/\Delta x \le M.$$

Since f is continuous,

$$m \approx f(x) \approx M,$$

so

$$f(x) \approx \Delta B/\Delta x,$$

$$f(x)\Delta x \approx \Delta B \text{ (compared to } \Delta x).$$

Then by the Infinite Sum Theorem,

$$B(a,b) = \int_a^b f(x)dx. \qquad \dashv$$

The Infinite Sum Theorem can often be used to show that there is only one reasonable way to define an intuitive concept as an integral, and also helps in discovering the correct integration formula. We now consider the area of a region. In Elementary Calculus we concentrated on a very simple type of region, the region between two continuous curves.

DEFINITION

A basic region in the (x,y) plane is a set of points of the form

$$\{(x,y) \in R^2: a \le x \le b,\ f(x) \le y \le g(x)\}$$

where a,b are real, f,g are real functions continuous on $[a,b]$, and $f(x) \le g(x)$ on $[a,b]$. It is called the region between f and g over $[a,b]$. The area of a basic region is defined as the integral

$$A = \int_a^b (g(x) - f(x))dx.$$

Justification Let $A(u,v)$ be the intuitive area between f and g over $[u,v]$. We shall prove the area formula from the following three assumptions:

$A(u,v)$ has the Addition Property. (4)

Subset Property. If D and E are basic regions and $D \subseteq E$, the area of D is at most the area of E. (5)

The area of a vertical rectangle is the product of the base and the height. (6)

Consider an infinitesimal subinterval $[x,x + \Delta x]^*$ of $[a,b]^*$.

We shall use (5) and (6) for hyperreal regions over $[x,x + \Delta x]^*$. The region between f and g over $[x,x + \Delta x]^*$ is a vertical strip of area $\Delta A = A(x,x + \Delta x)$. By the Hyperreal Extreme Value Theorem, f and g have minimum values f_{min}, g_{min} and maximum values f_{max}, g_{max} in $[x,x + \Delta x]^*$. The strip ΔA has an inscribed rectangle of height $g_{min} - f_{max}$ and a circumscribed rectangle of height $g_{max} - f_{min}$. Both rectangles have width Δx. By (5) and (6),

$$(g_{min} - f_{max})\Delta x \le \Delta A \le (g_{max} - f_{min})\Delta x,$$

$$g_{min} - f_{max} \le \Delta A/\Delta x \le g_{max} - f_{min} .$$

Since f and g are continuous, we have

$$f_{min} \approx f(x) \approx f_{max},\ g_{min} \approx g(x) \approx g_{max} .$$

Thus

$$g_{min} - f_{max} \approx g(x) - f(x) \approx g_{max} - f_{min},$$

$$\Delta A/\Delta x \approx g(x) - f(x),$$
$$\Delta A \approx (g(x) - f(x))\Delta x \text{ (compared to } \Delta x).$$

By the Infinite Sum Theorem,

$$A(a,b) = \int_a^b (g(x) - f(x))dx .$$ ⊣

We shall next use the Infinite Sum Theorem to justify the formulas for volumes of solids of revolution.

DEFINITION

Let D be a basic region in the plane of the form

$$D = \{(x,y): a \le x \le b,\ 0 \le y \le g(x)\}.$$

(i) The solid formed by revolving D about the x axis has volume

$$V = \int_a^b \pi(g(x))^2 dx.$$

(ii) The solid formed by revolving D about the y axis has volume

$$V = \int_a^b 2\pi\, x\, g(x)\, dx\ .$$

Justification (i) We use the disc method. Let $V(u,w)$ be the volume of the solid generated around the x axis by the region between 0 and g over $[u,w]$. We assume:

$V(u,w)$ has the Addition Property. (7)

Subset Property. If S and T are solids of revolution and $S \subseteq T$, the volume of S is at most the volume of T. (8)

A right circular cylinder with base of radius r and thickness h has volume $\pi r^2 h$. (9)

We use (8) and (9) for solids generated by regions over an infinitesimal subinterval $[x, x + \Delta x]^*$ of $[a,b]^*$. The region between 0 and g over $[x, x + \Delta x]^*$ generates an infinitely thin disc of volume ΔV. g has minimum value m and maximum value M in $[x, x + \Delta x]^*$, so ΔV has an inscribed cylinder of radius m and a circumscribed cylinder of radius M. Both cylinders have thickness Δx, so by (8) and (9),

$$\pi m^2 \Delta x \le \Delta V \le \pi M^2 \Delta x.$$

By continuity, $m \approx g(x) \approx M$, so

$$\Delta V \approx \pi (g(x))^2 \Delta x \ \text{(compared to } \Delta x).$$

The Infinite Sum Theorem gives

$$V(a,b) = \int_a^b \pi (g(x))^2 dx\ .$$

(ii) We use the cylindrical shell method. This time $V(u,w)$ denotes the volume generated about the y axis by the region between 0 and g over $[u,w]$. We assume the same properties (7) - (9). The region between 0 and g over $[x, x + \Delta x]^*$ generates a volume ΔV shaped like an infinitely thin cylindrical shell. By (7) and (9), the inscribed cylindrical shell has volume

$$\pi (x + \Delta x)^2 m - \pi x^2 m = \pi (2x + \Delta x) \Delta x\, m.$$

Using a similar formula for the circumscribed shell and (8), we have

$$\pi(2x + \Delta x)\Delta x\ m \le \Delta V \le \pi(2x + \Delta x)\Delta x\ M,$$

$$\pi(2x + \Delta x)m \le \Delta V/\Delta x \le \pi(2x + \Delta x)M.$$

By continuity, $m \approx g(x) \approx M$, so

$$\Delta V/\Delta x \approx \pi(2x + \Delta x)g(x) \approx 2\pi\ x\ g(x),$$

$$\Delta V \approx 2\pi\ x\ g(x)\ \Delta x \text{ (compared to } \Delta x).$$

The Infinite Sum Theorem gives

$$V(a,b) = \int_a^b 2\pi\ x\ g(x)\ dx\ . \qquad \dashv$$

We give just one example from physics here. Other examples are sketched in Elementary Calculus, and the detailed use of the Infinite Sum Theorem can readily be worked out.

DEFINITION

Suppose a plane object occupies the basic region

$$a \le x \le b,\ 0 \le y \le g(x)$$

and its density (mass per unit area) at (x,y) is a continuous function $\rho(x)$ of x alone. The mass of the object is

$$\text{mass} = \int_a^b g(x)\ \rho(x)\ dx\ .$$

Justification Let $m(u,w)$ be the mass of the piece of the object from $x = u$ to $x = v$. Our assumptions are:

$m(u,w)$ has the Addition Property. (10)

If the region of a plane object S is a subset of the region of T and the density of S is everywhere at most the density of T, then the mass of S is at most the mass of T. (11)

The mass of a vertical rectangular object of constant density is the product of the base, the height, and the density. (12)

On an infinitesimal subinterval $[x, x + \Delta x]^*$ of $[a,b]^*$, the continuous functions g and ρ have minimum values g_{min}, ρ_{min} and maximum values g_{max}, ρ_{max}. Δm is an infinitely thin strip of width Δx. Using (11) and (12),

$$g_{min}\rho_{min}\Delta x \leq \Delta m \leq g_{max}\rho_{max}\Delta x \; .$$

By continuity, $g_{min} \approx g(x) \approx g_{max}$, $\rho_{min} \approx \rho(x) \approx \rho_{max}$, and hence

$$\Delta m \approx g(x)\ \rho(x)\ \Delta x \quad \text{(compared to } \Delta x).$$

Applying the Infinite Sum Theorem,

$$m(a,b) = \int_a^b g(x)\ \rho(x)\ dx \; . \qquad \dashv$$

6B LENGTHS OF CURVES (§6.3, §6.4)

DEFINITION

A real curve

$$y = f(x),\ a \leq x \leq b$$

is said to be smooth if the derivative dy/dx is continuous on $[a,b]$. The length of a smooth curve is defined as

$$s = \int_a^b \sqrt{1 + \left(\frac{dy}{dx}\right)^2}\ dx \; .$$

Justification Let $s(u,w)$ be the intuitive length of the curve segment over $[u,w]$. Our assumptions are:

$s(u,w)$ has the Addition Property. (1)

If the slope of a smooth curve is infinitely close to the slope of the chord from $P(x,y)$ to $Q(x + \Delta x,\ y + \Delta y)$ at every point of $[x, x + \Delta x]^*$, then

$$\frac{\Delta s}{\sqrt{\Delta x^2 + \Delta y^2}} \approx 1 .$$

Here Δs is the length of the curve segment over $[x, x + \Delta x]^*$ and $\sqrt{\Delta x^2 + \Delta y^2}$ is the length of the chord. (2)

On an infinitesimal subinterval $[x, x + \Delta x]^*$, the infinitesimal curve segment Δs connects $P(x,y)$ to $Q(x + \Delta x, y + \Delta y)$. The chord from P to Q has slope $\Delta y/\Delta x$. It follows from Corollary 4 of Section 3C on continuous derivatives that $f'(x) \approx \Delta y/\Delta x$. Since f' is continuous, $f'(u) \approx \Delta y/\Delta x$ for all $u \in [x, x + \Delta x]^*$. By (2) and (3),

$$\frac{\Delta s}{\sqrt{\Delta x^2 + \Delta y^2}} \approx 1 .$$

Then

$$\frac{\Delta s}{\Delta x} = \frac{\Delta s}{\sqrt{\Delta x^2 + \Delta y^2}} \cdot \frac{\sqrt{\Delta x^2 + \Delta y^2}}{\Delta x}$$

$$= \frac{\Delta s}{\sqrt{\Delta x^2 + \Delta y^2}} \sqrt{1 + \left(\frac{\Delta y}{\Delta x}\right)^2} \approx \sqrt{1 + \left(\frac{\Delta y}{\Delta x}\right)^2} .$$

Since $dy/dx \approx \Delta y/\Delta x$,

$$\frac{\Delta s}{\Delta x} \approx \sqrt{1 + \left(\frac{\Delta y}{\Delta x}\right)^2} \approx \sqrt{1 + \left(\frac{dy}{dx}\right)^2} ,$$

$$\Delta s \approx \sqrt{1 + \left(\frac{dy}{dx}\right)^2} \, \Delta x \quad \text{(compared to } \Delta x\text{)}.$$

By the Infinite Sum Theorem,

$$s(a,b) = \int_a^b \sqrt{1 + \left(\frac{dy}{dx}\right)^2} \, dx . \qquad \dashv$$

As an application of the length formula we obtain the classical formula for the area of a sector of a circle.

THEOREM 1

On the circle $x^2 + y^2 = r^2$ of radius r, let P be the point (r,0) and let Q be a point (x,y) in the first quadrant. Then $A = \frac{1}{2}rs$ where A is the area of the sector POQ, r is the radius, and s is the length of the arc PQ.

Proof Since the circle is vertical at P, we shall take y as the independent variable instead of x. In the first quadrant the circle has the equation $x = \sqrt{r^2 - y^2}$. Then

$$s = \int_0^y \sqrt{1 + (\frac{dx}{dy})^2}\, dy ,$$

$$\frac{dx}{dy} = - \frac{y}{\sqrt{r^2 - y^2}} = - \frac{y}{x} ,$$

$$\sqrt{1 + (\frac{dx}{dy})^2} = \sqrt{1 + \frac{y^2}{x^2}} = \frac{\sqrt{x^2 + y^2}}{x} = \frac{r}{x} .$$

Let A be the area of the sector POQ,

$$A = \int_0^y \sqrt{r^2 - u^2}\, du - \frac{1}{2} x y .$$

Letting y vary, we have

$$\frac{dA}{dy} = \sqrt{r^2 - y^2} - \frac{1}{2}(x + y \frac{dx}{dy})$$

$$= x - \frac{1}{2} x - \frac{1}{2} y(- \frac{y}{x}) = \frac{x^2 + y^2}{2x} = \frac{r^2}{2x} .$$

Therefore

$$\frac{dA}{dy} = \frac{1}{2} r\sqrt{1 + (\frac{dx}{dy})^2} .$$

Integrating from 0 to y we obtain

$$A = \frac{1}{2} r\int_0^y \sqrt{1 + (\frac{dx}{dy})^2}\, dy = \frac{1}{2} r s .$$ ⊣

If we define π as the length of the unit semicircle, we see on setting $r = 1$ that the unit circle has circumference 2π and therefore has area π.

We next consider the length of a parametric curve.

DEFINITION

A parametric curve

$$x = f(t),\ y = g(t),\ a \le t \le b$$

is said to be <u>smooth</u> if at every point $t \in [a,b]$, f' and g' are continuous and at least one of $f'(t)$, $g'(t)$ is nonzero. The <u>path</u> of the parametric curve is the set

$$\{(f(t),\ g(t)):\ a \le t \le b\}.$$

The <u>length</u> of a smooth parametric curve is

$$s = \int_a^b \sqrt{\left(\frac{dx}{dt}\right)^2 + \left(\frac{dy}{dt}\right)^2}\, dt\ .$$

The justification of the parametric length formula is exactly like the justification of the length formula for curves of the form $y = f(x)$. Notice that when $x = t$, the parametric equation and length formula reduce to the special case

$$y = g(x),\ a \le x \le b,\ s = \int_a^b \sqrt{1 + \left(\frac{dy}{dx}\right)^2}\, dx\ .$$

We now show that the length of a simple parametric curve depends only on its path.

DEFINITION

A smooth parametric curve

$$C:\ a \le t \le b,\ x = f(t),\ y = g(t)$$

is said to be <u>simple</u> if C maps $[a,b]$ one to one onto its path and there is no t with $f'(t) = g'(t) = 0$. C_1 is a <u>reparametrization</u> of C if C and C_1 are simple parametric curves with the same path P.

REPARAMETRIZATION THEOREM

Let

$$C: a \le t \le b,\ x = f(t),\ y = g(t)$$

be a simple parametric curve.

(i) A parametric curve

$$C_1: A \le u \le B,\ x = F(u),\ y = G(u)$$

is a reparametrization of C if and only if there is a smooth function $t = h(u)$ of $[A,B]$ onto $[a,b]$ such that $h'(u)$ is never zero and for all u,

$$F(u) = f(h(u)),\ G(u) = g(h(u)).$$

(ii) If C_1 is a reparametrization of C then C and C_1 have the same length.

Proof (i) Suppose C_1 is a reparametrization of C. There are two cases for the endpoints.

Case 1. $F(A) = f(a)$, $F(B) = f(b)$.

Case 2. $F(A) = f(b)$, $F(B) = f(a)$.

We give the proof in Case 1. Since C and C_1 are one to one functions onto the same path P, there is a function h mapping $[A,B]$ onto $[a,b]$ such that whenever $t = h(u)$,

$$(F(u), G(u)) = (f(t), g(t)).$$

In Case 1, h is increasing. We show that h is smooth and $h'(u)$ is never zero. Let $A \le u < u + \Delta u \le B$ with $\Delta u \approx 0$, and let Δx, Δy, and Δt be the corresponding changes in $x = F(u)$, $y = G(u)$, $t = h(u)$. Then $\sqrt{\Delta x^2 + \Delta y^2}$ and hence Δt are positive infinitesimals, so h is continuous. We have

$$\frac{\Delta t}{\Delta u} = \frac{\sqrt{\Delta x^2 + \Delta y^2}/\Delta u}{\sqrt{\Delta x^2 + \Delta y^2}/\Delta t} = \frac{\sqrt{(\Delta x/\Delta u)^2 + (\Delta y/\Delta u)^2}}{\sqrt{(\Delta x/\Delta t)^2 + (\Delta y/\Delta t)^2}}$$

$$\approx \frac{\sqrt{F'(st(u))^2 + G'(st(u))^2}}{\sqrt{f'(st(t))^2 + g'(st(t))^2}},$$

so for real u, $h'(u)$ exists and

$$h'(u) = \sqrt{F'(u)^2 + G'(u)^2} \Big/ \sqrt{f'(h(u))^2 + g'(h(u))^2} \ .$$

Since h, F', G', f', and g' are continuous, the derivative h' is continuous.

We now prove the converse. Since h' is never zero, h maps [A,B] one to one onto [a,b]. Therefore C_1 maps [A,B] one to one onto the path of C. By the Chain Rule,

$$F'(u) = f'(h(u))h'(u), \ G'(u) = g'(h(u))h'(u).$$

Thus F' and G' are continuous and are never both zero, so C_1 is a simple parametric curve.

(ii) We give the proof in Case 1, $F(A) = f(a)$ and $G(A) = g(a)$. Let h be as in (i). Then $h'(u) > 0$, so

$$\text{length of } C = \int_a^b \sqrt{f'(t)^2 + g'(t)^2}\, dt$$

$$= \int_A^B \sqrt{f'(h(u))^2 + g'(h(u))^2}\, h'(u)du$$

$$= \int_A^B \sqrt{F'(u)^2 + G'(u)^2}\, du = \text{length of } C_1. \quad \dashv$$

We now justify the formula for the area of a surface of revolution. Our starting point is the elementary formula

$$A = \pi(r_1 + r_2)\ell$$

for the surface area of a circular cone frustum of slant height ℓ and bases of radius r_1 and r_2.

DEFINITION

The <u>area</u> of the surface formed by revolving a smooth curve

$$y = f(x), \ a \le x \le b$$

about the y axis is

$$A = \int_a^b 2\pi\, x\sqrt{1 + (\frac{dy}{dx})^2}\, dx\ .$$

The area of the surface of rotation about the x axis is

$$A = \int_a^b 2\pi\, y\sqrt{1 + (\frac{dy}{dx})^2}\, dx\ .$$

Justification We justify the area formula for rotation about the y axis. Let $s(u,v)$ and $A(u,v)$ be the length of the curve and the area of the surface of revolution for $u \le x \le v$. Assume $A(u,v)$ has the Addition Property. Let $[x,x + \Delta x]^*$ be an infinitesimal subinterval of $[a,b]^*$. When the line segment from (x,y) to $(x + \Delta x,\ y + \Delta y)$ is rotated about the y axis it generates an infinitely thin cone frustum of slant height $\sqrt{\Delta x^2 + \Delta y^2}$ and bases of radius x and $x + \Delta x$. This frustum has surface area

$$\pi(x + (x + \Delta x))\sqrt{\Delta x^2 + \Delta y^2}\ .$$

The curve $y = f(x)$ has value and slope infinitely close to the line segment from (x,y) to $(x + \Delta x,\ y + \Delta y)$ throughout the interval $[x,x + \Delta x]^*$. It is thus reasonable to assume that the surface area $\Delta A = A(x,x + \Delta x)$ and the frustum area have a ratio infinitely close to one.

$$\frac{\Delta A}{\pi(x + (x + \Delta x))\sqrt{\Delta x^2 + \Delta y^2}} \approx 1\ .$$

Since $\Delta y/\Delta x \approx dy/dx$,

$$\frac{\Delta A}{\pi(x + (x + \Delta x))\sqrt{\Delta x^2 + \Delta y^2}} \approx \frac{\Delta A}{2\pi\, x\sqrt{\Delta x^2 + \Delta y^2}}$$

$$= \frac{\Delta A/\Delta x}{2\pi\, x\sqrt{1 + (\frac{\Delta y}{\Delta x})^2}} \approx \frac{\Delta A/\Delta x}{2\pi\, x\sqrt{1 + (\frac{dy}{dx})^2}}\ .$$

Therefore

$$\Delta A \approx 2\pi\, x\sqrt{1 + (\frac{dy}{dx})^2}\,\Delta x \quad \text{(compared to } \Delta x\text{)}\ ,$$

and by the Infinite Sum Theorem,

$$A(a,b) = \int_a^b 2\pi\, x\sqrt{1 + \left(\frac{dy}{dx}\right)^2}\, dx\ .$$

The justification of the area formula for rotation about the x axis is similar. ⊣

Given a simple parametric curve

$$x = f(t),\ y = g(t),\ a \le t \le b,$$

the area of the surface of revolution about the y axis is defined by

$$A = \int_a^b 2\pi\, x\sqrt{\left(\frac{dx}{dt}\right)^2 + \left(\frac{dy}{dt}\right)^2}\, dt\ .$$

6C IMPROPER INTEGRALS (§6.7)

In Chapter 4 we only considered integrals of continuous functions over closed intervals. Using improper integrals we can integrate functions with at most finitely many discontinuities over arbitrary intervals.

DEFINITION

Suppose f is continuous on the half-open interval [a,b). The <u>improper integral</u> of f from a to b is defined as the limit

$$\int_a^b f(x)\,dx = \lim_{u \to b^-} \int_a^u f(x)\,dx\ .$$

If the limit exists the integral is said to <u>converge</u>. Otherwise the integral is said to <u>diverge</u>. If the limit is ∞ we say that the integral <u>diverges to</u> ∞ and write

$$\int_a^b f(x)\,dx = \infty.$$

Other types of improper integrals are defined analogously. For example: If f is continuous on (a,b],

$$\int_a^b f(x)dx = \lim_{u \to a^+} \int_u^b f(x)dx .$$

If f is continuous on $[a,\infty)$,

$$\int_a^\infty f(x)dx = \lim_{u \to \infty} \int_a^u f(x)dx .$$

The definitions can be rephrased using the infinitesimal definition of limit. We state four cases. Let f be continuous on [a,b).

1. $\int_a^b f(x)dx = L$ if and only if whenever $u < b$ but $u \approx b$, $\int_a^u f(x)dx \approx L$.
2. $\int_a^b f(x)dx = \infty$ if and only if whenever $u < b$ but $u \approx b$, $\int_a^u f(x)dx$ is positive infinite.

Let f be continuous on $[a,\infty)$.

3. $\int_a^\infty f(x)dx = L$ if and only if for every positive infinite H, $\int_a^H f(x)dx \approx L$.
4. $\int_a^\infty f(x)dx = \infty$ if and only if for every positive infinite H, $\int_a^H f(x)dx$ is positive infinite.

THEOREM 1

If f is continuous on [a,b], then the improper integral of f from a to b converges and equals the definite integral.

<u>Proof</u> Let $F(u) = \int_a^u f(x)dx$. By Section 4C, F is continuous on [a,b]. Therefore

$$\int_a^b f(x)dx = \lim_{u \to b^-} \int_a^u f(x)dx$$

where $\int_a^b f(x)dx$ denotes the definite integral. ⊣

THEOREM 2

Let f be continuous and nonnegative for x in $[a,\infty)$. Then the improper integral $\int_a^\infty f(x)dx$ either converges or diverges to ∞.

<u>Proof</u> Let $F(u) = \int_a^u f(x)dx$. Then F is continuous and nondecreasing on $[a,\infty)$.

Case 1. The range of F is bounded. Then it has a least upper bound L. Let H be positive infinite. Since every real $u \geq a$ is a solution of $F(u) \leq L$, we have $F(H) \leq L$. Consider a real number $r < L$. There is a real $u \geq a$ such that $F(u) > r$. Any real solution of $v \geq u$ is a solution of $F(v) \geq F(u)$ and hence $F(v) > r$. Therefore $F(H) > r$. Since this holds for all real $r < L$, we conclude that $F(H) \approx L$, and $\int_a^\infty f(x)dx = L$.

Case 2. The range of F is not bounded. Again let H be positive infinite. For every real r there is a real $u \geq a$ such that $F(u) > r$. Since $F(H) \geq F(u)$ we have $F(H) > r$ and hence $F(H)$ is positive infinite. Thus $\int_a^\infty f(x)dx = \infty$. ⊣

7 TRIGONOMETRIC FUNCTIONS

The starting point in the study of trigonometric functions is the notion of arc length on the unit circle. In Elementary Calculus we gave a geometric treatment based on the formula s = 2A for arc length on the unit circle where A is the area of the sector subtended by the arc s. That treatment does not depend on the integration formula for curve length and is familiar to students from pre-calculus courses.

Here we present a more concise analytic treatment which depends on the integration formula for curve length and begins with the Inverse Function Theorem.

7A INVERSE FUNCTION THEOREM (§7.3)

Given a binary relation $X \subseteq R^2$ on the reals, the inverse relation of X is the relation

$$Y = \{(y,x) : (x,y) \in X\}.$$

Obviously, if Y is the inverse relation of X then X is the inverse relation of Y. If f and g are real functions and g is the inverse relation of f, we call g the inverse function of f. Here are some simple facts about inverse functions.

LEMMA 1 (i) A real function f has an inverse function if and only if f is one to one.

(ii) g is the inverse function of f if and only if the equations

$$f(x) = y, \quad g(y) = x$$

have the same solutions.

(iii) g is the inverse function of f if and only if

$$\text{domain}(g) = \text{range}(f)$$

and $g(f(x)) = x$ for all $x \in \text{domain}(f)$.

We remark that if f is increasing,

$$x < y \text{ implies } f(x) < f(y),$$

then f is one to one and therefore has an inverse function. Similarly, if f is decreasing then f has an inverse function.

THEOREM 2

Let f be continuous and increasing on its domain which is an interval I, and let g be the inverse function of f. Then g is increasing, the domain of g is an interval J, and g is continuous on J.

Proof Let J be the domain of g. Suppose $A,B \in J$ and $A < B$. Let $a = g(A)$, $b = g(B)$. Then $f(a) = A$, $f(b) = B$. We cannot have $b \leq a$ because that would imply $f(b) \leq f(a)$, $B \leq A$. Therefore $a < b$, so g is increasing.

Suppose C is a real number with $A < C < B$. By the Intermediate Value Theorem for the continuous function f, there is a point $c \in (a,b)$ such that $f(c) = C$. Therefore C belongs to the range of f, which is the domain J of g. This shows that J is an interval. To show that g is continuous let $A \in J$, $X \in J^*$, $A \approx X$. Put $a = g(A)$, $x = g(X)$. We must show that $a \approx x$. Assume first that $A < X$. We have $a \in I$ and $f(a) = A$. By the Solution Axiom we also have $x \in I^*$ and $f(x) = X$. Also, since g is increasing, g^* is increasing, so $a < x$. If $a \not\approx x$, then there is a real number $c \in I$ such that $a < c < x$, hence $f(a) < f(c) < f(x)$, contradicting $f(a) = A \approx X = f(x)$. This shows that $a \approx x$. The case $X > A$ is similar, so g is continuous on J. ⊣

INVERSE FUNCTION THEOREM

Suppose f is continuous and increasing on an open interval I, and g is the inverse function of f. For any point $x \in I$ where f is differentiable and $f'(x) \neq 0$, g is differentiable at $y = f(x)$ and $g'(y) = 1/f'(x)$.

<u>Proof</u> Let Δy be a nonzero infinitesimal. We must show that

$$\frac{g(y + \Delta y) - g(y)}{\Delta y} \approx \frac{1}{f'(x)}.$$

Let J be the range of f and the domain of g. Since I is an open interval, J is an open interval. Thus $y = f(x)$ is an interior point of J. Therefore $g(y + \Delta y)$ is defined. We have $x = g(y)$, and we put

$$\Delta x = g(y + \Delta y) - g(y).$$

By Theorem 2, g is continuous and increasing, so Δx is infinitesimal and not zero. Also,

$$x + \Delta x = g(y + \Delta y), \; f(x + \Delta x) = y + \Delta y.$$

Then

$$f'(x) \approx \frac{f(x + \Delta x) - f(x)}{\Delta x} = \frac{\Delta y}{\Delta x},$$

and since $f'(x) \neq 0$,

$$\frac{1}{f'(x)} \approx \frac{\Delta x}{\Delta y} = \frac{g(y + \Delta y) - g(y)}{\Delta y}.$$ ⊣

We shall now prove another Inverse Function Theorem, due to Behrens (see the book of Stroyan). It does not require the hypothesis that f is continuous on a neighborhood, and will be convenient when we study two variables later.

DEFINITION

A real function f is said to be <u>uniformly differentiable</u> at a real point c if $f'(c)$ exists and whenever $x \approx c$ and Δx is nonzero infinitesimal,

$$f'(c) \approx \frac{f(x + \Delta x) - f(x)}{\Delta x}.$$

Theorem 3 of Section 3C shows that f has a continuous derivative on an open interval I if and only if f is uniformly differentiable at every real point $c \in I$. We now consider uniform differentiability at a single real point.

THEOREM 3

If f is uniformly differentiable at c, then f is continuous on some real neighborhood I of c.

Proof Let $L = |f'(c)| + 1$. It suffices to prove that on some real neighborhood $(c - r, c + r)$ of c, we have the Lipschitz condition

$$|f(y) - f(x)| \le L|y - x|, \text{ all } x,y \in (c - r, c + r). \quad (1)$$

f is defined on some real neighborhood of c. Suppose there is no real number $r > 0$ such that (1) holds. Then every real solution of $r > 0$ is a partial solution of

$$|f(y) - f(x)| > L|y - x|, \; x,y \in (c - r, c + r). \quad (2)$$

Let r_1 be a positive infinitesimal. By the Partial Solution Theorem there are hyperreal x_1, y_1 such that (2) holds. But then

$$\left|\frac{f(y_1) - f(x_1)}{y_1 - x_1}\right| > |f'(c)| + 1, \; x_1 \approx c, \; y_1 \approx c,$$

contradicting the uniform differentiability of f at c. ⊣

LOCAL INVERSE FUNCTION THEOREM

Suppose f is increasing, f is uniformly differentiable at a real point x, and $f'(x) \neq 0$. Then the inverse function g is uniformly differentiable at $y = f(x)$ and $g'(y) = 1/f'(x)$.

Proof By Theorem 3, f is continuous on some real neighborhood I of c. Then by the previous Inverse Function Theorem, $g'(y)$ exists and $g'(y) = 1/f'(x)$. ⊣

The following application of the Inverse Function Theorem shows that in a simple parametric curve we may take the curve length itself as the independent variable.

THEOREM 4

Let

$$C:\ a \le t \le b,\ x = f(t),\ y = g(t)$$

be a simple parametric curve. Then C has a reparametrization

$$C_1:\ 0 \le s \le L,\ x = F(s),\ y = G(s),$$

such that s is equal to the length of the curve from 0 to s.

Proof Let $\ell(t)$ be the curve length from A to t,

$$\ell(t) = \int_A^t \sqrt{f'(u)^2 + g'(u)^2}\, du .$$

Then $\ell'(t) = \sqrt{f'(t)^2 + g'(t)^2}$ is always positive and continuous, and ℓ maps $[a,b]$ onto $[0,L]$ where L is the length of C. By the Inverse Function Theorem the inverse function h of ℓ maps $[0,L]$ onto $[a,b]$ and has the derivative $h'(s) = 1/\ell'(h(s)) > 0$. Since h and ℓ' are continuous, h' is continuous. By the Reparametrization Theorem in Section 6B, the curve

$$C_1:\ 0 \le s \le L,\ x = f(h(s)),\ y = g(h(s))$$

is a reparametrization of C, and the length of C_1 from 0 to s is $\ell(h(s)) = s$. ⊣

7B DERIVATIVES OF TRIGONOMETRIC FUNCTIONS (§7.1, §7.2)

We first define the trigonometric functions and then use the Inverse Function Theorem to show that they are differentiable.

DEFINITION

π is defined as the length of the unit semicircle $x = \sqrt{1 - y^2}$, $-1 \le y \le 1$.

DEFINITION

Let $0 \le \theta < \pi/2$ and let $P(x,y)$ be the point at distance θ counter-clockwise around the unit circle starting from $(1,0)$. Then we define

$$x = \cos \theta, \ y = \sin \theta.$$

LEMMA 1 For $0 \le \theta < \pi/2$, the functions $y = \sin \theta$ and $x = \cos \theta$ are differentiable and

$$\frac{d(\sin \theta)}{d\theta} = \cos \theta, \ \frac{d(\cos \theta)}{d\theta} = \sin \theta.$$

Proof We repeat the argument used for Theorem 4 of the preceding section but for the special case $x = \sqrt{1 - y^2}$, $0 \le y < 1$.

We have

$$\frac{dx}{dy} = \frac{-y}{\sqrt{1 - y^2}}$$

$$\sqrt{1 + \left(\frac{dx}{dy}\right)^2} = \sqrt{1 + \frac{y^2}{1 - y^2}} = \frac{1}{\sqrt{1 - y^2}} .$$

Let θ be the arc length

$$\theta = \int_0^y \frac{1}{\sqrt{1 - u^2}}\, du .$$

By definition, $x = \cos \theta$ and $y = \sin \theta$. For $0 \le y < 1$, θ is increasing and has the continuous derivative

$$\frac{d\theta}{dy} = \frac{1}{\sqrt{1 - y^2}} .$$

By the Inverse Function Theorem, $y = \sin \theta$ is differentiable and

$$\frac{dy}{d\theta} = \frac{1}{d\theta/dy} = \sqrt{1 - y^2} = x = \cos\theta.$$

Similarly,

$$\frac{dx}{d\theta} = \frac{dx}{dy}\frac{dy}{d\theta} = -y = -\sin\theta. \qquad \dashv$$

We extend the sine and cosine functions from the interval $[0,\pi/2)$ to the whole real line by defining

$$\sin(\theta + \pi/2) = \cos\theta,$$

$$\sin(\theta + \pi) = -\sin\theta.$$

One can easily check that the sine and cosine have period 2π. Given the derivative formulas for $0 \le \theta < \pi/2$ in Lemma 1, we leave the proof of the formulas for arbitrary θ to the reader.

THEOREM 2

$$\frac{d(\sin\theta)}{d\theta} = \cos\theta, \quad \frac{d(\cos\theta)}{d\theta} = -\sin\theta.$$

The other trigonometric functions are defined by

$$\tan\theta = \frac{\sin\theta}{\cos\theta}, \quad \cot\theta = \frac{\cos\theta}{\sin\theta},$$

$$\sec\theta = \frac{1}{\cos\theta}, \quad \csc\theta = \frac{1}{\sin\theta}.$$

The inverse trigonometric functions are obtained by restricting the trigonometric functions to either $[-\pi/2,\pi/2]$ or $[0,\pi]$ and then taking the inverse functions. For example,

arc sin x is the inverse of $\sin\theta$, $-\pi/2 \le \theta \le \pi/2$,

arc tan x is the inverse of $\tan\theta$, $-\pi/2 \le \theta \le \pi/2$,

arc sec x is the inverse of $\sec\theta$, $0 \le \theta \le \pi$.

The Inverse Function Theorem shows that the inverse trigonometric functions are differentiable except at the endpoints and leads in the usual way to the formulas for the derivatives:

$$\frac{d(\text{arc sin } x)}{dx} = \frac{1}{\sqrt{1 - x^2}}, \quad |x| < 1 .$$

$$\frac{d(\text{arc tan } x)}{dx} = \frac{1}{1 + x^2}, \quad \text{all } x .$$

$$\frac{d(\text{arc sec } x)}{dx} = \frac{1}{|x| \sqrt{x^2 - 1}}, \quad |x| > 1 .$$

7C AREA IN POLAR COORDINATES (§7.9)

We assume the reader is familiar with the polar coordinate system. The Infinite Sum Theorem can be used to obtain a formula for area in polar coordinates.

DEFINITION

By a basic polar region we mean a set of points with polar coordinates of the form

$$\{(r,\theta): a \le \theta \le b, \ f(\theta) \le r \le g(\theta)\},$$

where $b \le a + 2\pi$, f and g are continuous on $[a,b]$, and $0 \le f(\theta) \le g(\theta)$.

Thus the image of a basic polar region under the mapping

$$(r,\theta) \to (r \cos \theta, \ r \sin \theta)$$

is a basic rectangular region. If the functions f and g are constant the region is called a polar rectangle. The simplest kind of polar rectangle is a circular sector

$$\{(r,\theta): a \le \theta \le b, \ 0 \le r \le c\}.$$

Our starting point for polar areas is the following formula for the area of a circular sector:

$$A = \frac{1}{2} c^2 (b - a).$$

This formula comes from the formula $A = \frac{1}{2} rs$ where r is the radius of the circle and $s = (b - a)r$ is the length of the arc. Consider a polar region of the simple form

$$D = \{(r,\theta): a \le \theta \le b,\ 0 \le r \le g(\theta)\}.$$

DEFINITION

Let $g(\theta)$ be a nonnegative continuous real function for $a \le \theta \le b$, where $b \le a + 2\pi$. By a polar area function for g we mean a function $A(u,v)$ defined for $u,v \in [a,b]$ with the following two properties.

(i) $A(u,w) = A(u,v) + A(v,w)$.

(ii) If g has minimum value m and maximum value M on $[u,v]$, then

$$\frac{1}{2} m^2(v - u) \le A(u,v) \le \frac{1}{2} M^2(v - u) .$$

Condition (ii) says that $A(u,v)$ is between the areas of the inscribed and circumscribed circular sectors for $u \le \theta \le v$.

THEOREM

The unique polar area function for g is the definite integral

$$A(a,b) = \int_a^b \frac{1}{2} g(\theta)^2 \, d\theta .$$

Proof $A(u,v)$ is a polar area function because

$$\int_u^v \frac{1}{2} m^2 d\theta \le \int_u^v \frac{1}{2} g(\theta)^2 d\theta \le \int_u^v \frac{1}{2} M^2 d\theta ,$$

$$\int_u^v \frac{1}{2} m^2 d\theta = \frac{1}{2} m^2(v - u), \quad \int_u^v \frac{1}{2} M^2 d\theta = \frac{1}{2} M^2(v - u).$$

To prove uniqueness let $B(u,v)$ be a polar area function for g. Let $[\theta,\theta + \Delta\theta]^*$ be an infinitesimal subinterval of $[a,b]^*$. On $[\theta,\theta + \Delta\theta]^*$, g has a minimum value m and a maximum value M. By continuity, $m \approx g(\theta) \approx M$. Applying the Solution Axiom to property (ii) we have

$$\frac{1}{2} m^2 \Delta\theta \le \Delta B \le \frac{1}{2} M^2 \Delta\theta,$$

$$\frac{1}{2} m^2 \le \Delta B/\Delta\theta \le \frac{1}{2} M^2,$$

$$\Delta B/\Delta\theta \approx \frac{1}{2} g(\theta)^2,$$

$$\Delta B \approx \frac{1}{2} g(\theta)^2 \Delta\theta \qquad \text{(compared to } \Delta\theta\text{).}$$

By (i), B has the Addition Property. Therefore by the Infinite Sum Theorem,

$$B(a,b) = \int_a^b \frac{1}{2} g(\theta)^2 d\theta. \qquad \dashv$$

The area formula for an arbitrary basic polar region, which is justified in a similar way, is

$$A = \int_a^b \frac{1}{2} (g(\theta)^2 - f(\theta)^2) d\theta.$$

8 EXPONENTIAL FUNCTIONS

The exponential and logarithmic functions have been introduced in a variety of ways in calculus texts. It seems to us that the most natural approach is to define a^x as the unique continuous function of x which has the value $a^{m/n} = \sqrt[n]{a^m}$ when x is rational. This is the approach used in Elementary Calculus. Here we shall use hyperreal numbers to prove a general result on uniquely extending continuous functions, and then apply the result to the case of the exponential functions.

8A EXTENDING CONTINUOUS FUNCTIONS

Let us recall the hyperreal characterizations of the closure of a set of reals (Section 1D) and of uniform continuity (Section 3C). The closure of a set $Y \subseteq R$ is the set

$$\overline{Y} = \{st(y): y \text{ is finite and } y \in Y^*\}.$$

A real function f is uniformly continuous on a set $Y \subseteq R$ if and only if for all $x, y \in Y^*$, $x \approx y$ implies $f(x) \approx f(y)$.

We shall see in the next section that the exponential function a^x is not uniformly continuous on the set Q of rationals but is uniformly continuous on every bounded subset of Q.

THEOREM 1

A real function f is uniformly continuous on every bounded subset of Y if and only if for every finite $x,y \in Y^*$, $x \approx y$ implies $f(x) \approx f(y)$.

<u>Proof</u> This follows at once from the fact that x, y are finite elements of Y^* if and only if $x,y \in (Y \cap [a,b])^*$ for some interval [a,b]. ⊣

THEOREM 2

Let f be a real function which is uniformly continuous on every bounded subset of its domain Y. Then f has a unique extension g whose domain is the closure $\overline{Y}$ of Y and which is continuous on $\overline{Y}$.

<u>Proof</u> Let g be the real function on $\overline{Y}$ defined by

$$g(st(x)) = st(f(x)) \text{ for finite } x \in Y^*.$$

This definition is unambiguous because if $st(x) = st(y)$ then $st(f(x)) = st(f(y))$. g extends f, because if $r \in Y$ then

$$g(r) = g(st(r)) = st(f(r)) = f(r) .$$

We show that g is continuous at each point $c \in \overline{Y}$. Let X be the bounded set $X = Y \cap [c - 1, c + 1]$. We use the ε,δ condition for uniform continuity in Section 5A. Consider a real $\varepsilon > 0$. There is a real δ, $1 > \delta > 0$, such that whenever $x,y \in X$ and $|x - y| < \delta$, $|f(x) - f(y)| < \varepsilon/2$. By the Solution Axiom, whenever $x,y \in X^*$ and $|x - y| < \delta$, $|f(x) - f(y)| < \varepsilon/2$. Now let $b \in \overline{Y}$ and $|b - c| < \delta$. We have $b = st(x)$, $c = st(y)$ for some $x,y \in Y^*$. Since $|b - c| < \delta < 1$ we have

$$x,y \in X^*, \quad |x - y| < \delta.$$

Therefore

$$|f(x) - f(y)| < \varepsilon/2$$

$$|st(f(x)) - st(f(y))| \le \varepsilon/2 < \varepsilon$$

$$|g(b) - g(c)| < \varepsilon .$$

Thus the ε,δ condition for continuity holds for g at c. The function g is unique because for any continuous extension h of f to $\overline{Y}$ and any point $c \in \overline{Y}$ we have

$$h(c) = h(st(x)) = st(h(x)) = st(f(x)) = g(c)$$

where $x \in Y^*$ and $c = st(x)$. ⊣

8B DEFINITION AND DERIVATIVE OF e^x (§8.1, §8.2)

In Elementary Calculus, hyperreal numbers were used as follows to define and obtain the basic properties of the exponential function. The natural extension of the set Q of rationals is called the set Q^* of hyperrationals. Here are some properties of Q^* which follow at once from the Solution Axiom.

THEOREM 1

(i) Q^* is a subfield of R^*, that is, Q^* is closed under addition, subtraction, multiplication, and division by nonzero numbers.

(ii) Q^* is dense in R^*, that is, if $b < c$ then there exists $q \in Q^*$ with $b < q < c$.

(iii) For every $x \in R^*$ there is a $q \in Q^*$ such that $x \approx q$. (This follows from (ii).)

The closure of the set Q of rationals is the whole real line R. One way to see this is by applying (iii) to $x \in R$. We shall use the following properties of rational exponents.

LEMMA 2 Let a, b be positive real numbers. If q, r are rational, then:

(i) $1^q = 1$

(ii) $a^{q+r} = a^q a^r$, $a^{q-r} = a^q/a^r$

(iii) $a^{qr} = (a^q)^r$

(iv) $a^q b^q = (ab)^q$

(v) $a < b$ and $q > 0$ implies $a^q < b^q$

(vi) $1 < a$ and $q < r$ implies $a^q < a^r$

(vii) $a \geq 0$ and $q \geq 1$ implies $(a + 1)^q \geq aq + 1$.

LEMMA 3 Let $a > 0$. The function $f(x) = a^x$, $x \in Q$, is uniformly continuous on every bounded subset of Q.

Proof Let $a \geq 1$. Suppose $q, r \in Q^*$, $q \approx r$, $q < r$. We must show that $a^q \approx a^r$. Let $b = a^{(r-q)} - 1$. By (vi) and the Solution Axiom, $b \geq 0$. Moreover, by (vii) and the Solution Axiom,

$$a = (b + 1)^{1/(r-q)} \geq \frac{b}{r - q} + 1 \geq 1 ,$$

so $b/(r - q)$ is finite and hence $b \approx 0$. Thus $a^{r-q} \approx 1$. For some integer n, $n \leq q < n + 1$, and by (vi), $a^n < a^q < a^{n+1}$. Therefore a^q is finite and $a^r \approx a^q$. When $0 < a \leq 1$ we have $a^{-1} \geq 1$, so

$$a^r = (a^{-1})^{-r} \approx (a^{-1})^{-q} = a^q . \qquad ⊣$$

By Theorem 2 in the preceding section we may make the following definition.

DEFINITION

Let a be a positive real number. Then the <u>exponential function with base</u> a, a^x, is the unique extension of the function a^q, $q \in Q$, which is continuous on the whole real line.

THEOREM 4

The exponent rules (i) - (vii) of Lemma 2 hold for arbitrary real numbers q and r.

Proof All the rules except (iii) are easily proved by using Lemma 2 and the fact that there are hyperrational numbers $q_1 \approx q$ and $r_1 \approx r$. We prove (iii) for the case $1 < a$, $q, r > 0$. Choose hyperrational numbers q_1, q_2 such that

$$q_1 \approx q_2,\ q_1 < q < q_2 .$$

Do the same for r. We have

$$q_1 r_1 < qr < q_2 r_2 .$$

By (vi) and the Solution Axiom,

$$a^{q_1 r_1} < a^{qr} < a^{q_2 r_2} ,$$

$$a^{q_1} < a^{q} < a^{q_2} ,$$

$$(a^{q_1})^{r_1} < (a^{q})^{r} < (a^{q_2})^{r_2} .$$

Using (iii) for hyperrational exponents,

$$a^{q_1 r_1} = (a^{q_1})^{r_1}, \quad a^{q_2 r_2} = (a^{q_2})^{r_2} .$$

But $q_1 r_1 \approx qr \approx q_2 r_2$, so by continuity, $a^{q_1 r_1} \approx a^{qr} \approx a^{q_2 r_2}$. It follows that $a^{qr} \approx (a^q)^r$, and since both numbers are real they must be equal. ⊣

DEFINITION

Let $0 < a$ and $a \neq 1$. The logarithmic function with base a is the inverse of the exponential function with base a,

$$x = \log_a y \quad \text{iff} \quad y = a^x .$$

By Theorem 2 in Section 7A, $\log_a y$ is continuous and has domain $(0,\infty)$. The familiar rules for logarithms follow from the corresponding rules for exponents and will be used freely below. We now introduce the number e.

LEMMA 5 The limit $\lim_{x\to\infty}(1 + \frac{1}{x})^x$ exists.

Proof Let b be any real number greater than 1. The function $y = b^t$ is continuous and positive, so $c = \int_0^1 b^t dt > 0$. Let H be positive infinite. We shall prove that

$$(1 + \frac{1}{H})^H \approx b^{c/(b-1)} .$$

We work with the logarithm

$$\log_b[(1 + \frac{1}{H})^H] = H \log_b(1 + \frac{1}{H}) .$$

Let

$$\Delta t = \log_b(1 + \frac{1}{H}) .$$

Δt is positive and infinitesimal, because

$$\Delta t \approx \log_b 1 = 0 .$$

Moreover,

$$b^{\Delta t} = 1 + \frac{1}{H}, \quad H = \frac{1}{b^{\Delta t} - 1},$$

$$H \log_b(1 + \frac{1}{H}) = \frac{\Delta t}{b^{\Delta t} - 1} .$$

We wish to estimate the Riemann sum $\Sigma_0^1 b^t \Delta t$. Any real solution of

$$\Delta u > 0, \ n \in Z, \ n\Delta u < 1 \le (n + 1)\Delta u \tag{1}$$

is a solution of

$$(1 + b^{\Delta u} + \dots + b^{(n-1)\Delta u})\Delta u \le \sum_0^1 b^u \Delta u$$

$$\le (1 + b^{\Delta u} + \dots + b^{n\Delta u})\Delta u .$$

By the geometric series formula, this simplifies to

$$\frac{b^{n\Delta u} - 1}{b^{\Delta u} - 1} \Delta u \le \sum_0^1 b^u \Delta u \le \frac{b^{(n+1)\Delta u} - 1}{b^{\Delta u} - 1} \Delta u . \tag{2}$$

Let K be the hyperinteger with $K\Delta t < 1 \le (K + 1)\Delta t$. By the Solution Axiom,

$$\frac{b^{K\Delta t} - 1}{b^{\Delta t} - 1} \Delta t \le \sum_0^1 b^t \Delta t \le \frac{b^{(K+1)\Delta t} - 1}{b^{\Delta t} - 1} \Delta t .$$

Then

$$H \log_b(1 + \frac{1}{H})(b^{K\Delta t} - 1) \le \sum_0^1 b^t \Delta t$$

$$\le H \log_b(1 + \frac{1}{H})(b^{(K+1)\Delta t} - 1) ,$$

$$\frac{\sum_0^1 b^t \Delta t}{b^{(K+1)\Delta t} - 1} \le H \log_b(1 + \frac{1}{H}) \le \frac{\sum_0^1 b^t \Delta t}{b^{K\Delta t} - 1} .$$

Since $K\Delta t \approx 1 \approx (K+1)\Delta t$, we conclude that

$$c/(b-1) \approx H \log_b(1 + \frac{1}{H}) \quad , \; c = \int_0^1 b^t dt$$

and

$$b^{c/(b-1)} \approx (1 + \frac{1}{H})^H .$$

Since this holds for all infinite H,

$$\lim_{x\to\infty} (1 + \frac{1}{x})^x = b^{c/(b-1)} .$$

DEFINITION

$$e = \lim_{x\to\infty} (1 + \frac{1}{x})^x \; , \text{ and } \ln x = \log_e x .$$

THEOREM 6

The exponential function e^x is differentiable, and

$$\frac{d(e^x)}{dx} = e^x .$$

<u>Proof</u> Let t be finite and Δt be positive infinitesimal. We show that

$$\frac{e^{t+\Delta t} - e^t}{\Delta t} \approx e^t .$$

We have

$$\frac{e^{t+\Delta t} - e^t}{\Delta t} = e^t \frac{e^{\Delta t} - 1}{\Delta t} .$$

Let $b = \dfrac{e^{\Delta t} - 1}{\Delta t}$. Then $e^{\Delta t} = 1 + b\Delta t$.
Since e^t is continuous, $e^{\Delta t} \approx 1$, and $b\Delta t$ is positive infinitesimal.
Thus $H = 1/(b\Delta t)$ is positive infinite. Also,

$$e \approx (1 + \frac{1}{H})^H = (1 + b\Delta t)^{1/b\Delta t} = (e^{\Delta t})^{1/b\Delta t} = e^{1/b} .$$

Therefore $b \approx 1$, and

$$\frac{e^{t+\Delta t} - e^t}{\Delta t} = be^t \approx e^t . \qquad (3)$$

Now let x be real and Δt positive infinitesimal. Then by (3),

$$\frac{e^{x+\Delta t} - e^x}{\Delta t} \approx e^x , \quad \frac{e^x - e^{x-\Delta t}}{\Delta t} \approx e^{x-\Delta t} \approx e^x .$$

Therefore $\frac{d(e^x)}{dx} = e^x$. ⊣

In general, $a^x = e^{x \ln a}$, so

$$\frac{d(a^x)}{dx} = \frac{d(e^{x \ln a})}{dx} = \ln a \; e^{x \ln a} = \ln a \; a^x .$$

This shows that e is uniquely characterized by the equation $d(e^x)/dx = e^x$, because when $a \neq e$, $\ln a \neq 1$.

COROLLARY 7 $y = \ln x$ is differentiable for all $x > 0$ and

$$\frac{d(\ln x)}{dx} = \frac{1}{x} .$$

Proof Since $y = \ln x$, $x = e^y$. By the Inverse Function Theorem, dy/dx exists and

$$\frac{dy}{dx} = \frac{1}{dx/dy} = \frac{1}{e^y} = \frac{1}{x} .$$ ⊣

9 INFINITE SERIES

Throughout this chapter we shall let H and K denote positive infinite hyperintegers. The set of natural numbers (nonnegative integers) will be denoted by N, and the set of positive integers by N^+.

9A SEQUENCES (§9.1)

By an <u>infinite sequence</u> $\langle a_n \rangle$ we mean a function from either N or N^+ into the real numbers.

DEFINITION

An infinite sequence $\langle a_n \rangle$ <u>converges</u> to a real number L if $a_H \approx L$ for all positive infinite hyperintegers H. In symbols, $\lim_{n\to\infty} a_n = L$. $\langle a_n \rangle$ <u>diverges</u> to ∞ if a_H is positive infinite for all positive infinite hyperintegers H. In symbols, $\lim_{n\to\infty} a_n = \infty$.

We begin with some computations of limits.

THEOREM 1

The following sequences diverge to ∞.

(i) $\lim_{n\to\infty} \frac{n!}{b^n} = \infty \quad (b \geq 1)$

(ii) $\lim_{n\to\infty} \frac{b^n}{n^c} = \infty \quad (b > 1,\ c \geq 0)$

(iii) $\lim_{n\to\infty} \frac{n^c}{\ln(n)} = \infty \quad (c > 0)$

(iv) $\lim_{n\to\infty} \ln(n) = \infty.$

Proof Let H be positive infinite. We show that each of

$$H!/b^H,\ b^H/H^c,\ H^c/\ln H,\ \ln H$$

are positive infinite. ln H is positive infinite because for each real r, $e^r < H$ and hence $r < \ln H$. In the other three cases we show that the logarithms are positive infinite.

(i) For an integer $m > b$ we have

$$\ln(\frac{H!}{b^H}) = \ln 1 + \ldots + \ln(m - 1) + \ln m + \ldots + \ln H - H \ln b$$

$$> (H - m)\ln m - H \ln b = H(\ln m - \ln b) - m \ln m .$$

Since $\ln m > \ln b$, $\ln(H!/b^H)$ is positive infinite.

(ii) We have

$$\ln(\frac{b^H}{H^c}) = H \ln b - c \ln H = H(\ln b - c \frac{\ln H}{H}) .$$

By l'Hospital's Rule, $\lim_{x\to\infty} (\ln x/x) = 0$, so $\ln H/H \approx 0$ and $\ln(b^H/H^c)$ is positive infinite.

(iii) Putting $K = \ln H$,

$$\ln(\frac{H^c}{\ln H}) = c \ln H - \ln(\ln H) = K(c - \frac{\ln K}{K}) ,$$

so $\ln(H^c/\ln H)$ is positive infinite. ⊣

Here are three equivalence theorems for limits of sequences.

THEOREM 2

Given a sequence $\langle a_n \rangle$ and a real number L, the following are equivalent.

(i) $\lim_{n\to\infty} a_n = L$.

(ii) There is an infinite H such that for all $K \geq H$, $a_K \approx L$.

(iii) The ε,M condition. For every real $\varepsilon > 0$ there is an integer M such that for all $n \geq M$, $|a_n - L| < \varepsilon$.

The proof is similar to the equivalence theorem for limits of functions in Section 5A.

THEOREM 3

The following are equivalent.

(i) $\lim_{n\to\infty} a_n = \infty$.

(ii) There is an infinite H such that for all $K \geq H$, a_K is positive infinite.

(iii) The ε,M condition. For every real B there exists an integer M such that for all $n \geq M$, $a_n > B$.

THEOREM 4

The following are equivalent.

(i) The sequence $\langle a_n \rangle$ converges.

(ii) Hyperreal Cauchy Condition. For all infinite H and K, $a_H \approx a_K$.

(iii) Real Cauchy Condition. For every real $\varepsilon > 0$ there is an integer M such that for all $m,n \geq M$, $|a_m - a_n| < \varepsilon$.

Proof Assume (i), say $\lim_{n\to\infty} a_n = L$. Then for all infinite H and K, $a_H \approx L \approx a_K$, so (ii) holds. Now assume (ii). Suppose the Real Cauchy Condition (iii) fails for some real $\varepsilon > 0$. Then every real $M \in N^+$ is a partial real solution of

$$m \in N^+,\ n \in N^+,\ M \leq m,\ M \leq n,\ |a_m - a_n| \geq \varepsilon.$$

Let M_1 be infinite. By the Partial Solution Theorem there exist infinite $H, K \geq M_1$ such that $|a_H - a_K| \geq \varepsilon$, contradicting (ii).

Assume the Real Cauchy Condition (iii). Let H be positive infinite. There is an integer M_1 such that for all integers $m,n \geq M_1$, $|a_m - a_n| < 1$. By the Solution Axiom, for all integers $m \geq M_1$,

$|a_m - a_H| < 1$. Therefore a_H is finite. Let $L = st(a_H)$. We show that the sequence converges to L. Let $\varepsilon > 0$ be real and let M be the corresponding integer. Using the Solution Axiom again, for all hyperreal $n \geq M$ we have $|a_n - a_H| < \varepsilon$. Thus for all infinite K, $|a_K - a_H| < \varepsilon$. Since this holds for all ε, we have $a_K \approx a_H \approx L$ for all infinite K, whence $\lim_{n\to\infty} a_n = L$. ⊣

We now give hyperreal proofs of the Bolzano-Weierstrass and the Countable Heine-Borel Theorems. A <u>subsequence</u> of $\langle a_n \rangle$ is a sequence $\langle a_{f(n)} \rangle$ where f is an increasing function from N into N.

THEOREM 5

Let H be positive infinite.

(i) If $st(a_H) = L$ then $\langle a_n \rangle$ has a subsequence converging to L.

(ii) If a_H is positive infinite then $\langle a_n \rangle$ has a subsequence diverging to ∞.

<u>Proof</u> (i) Each positive integer n is a partial hyperreal solution of

$$m \in N^+, \ m \geq n, \ |a_m - L| < \frac{1}{n}$$

and therefore a partial real solution. By induction define f(n) to be the first positive integer such that

$$f(n) > f(n - 1), \ |a_{f(n)} - L| < \frac{1}{n} .$$

By the ε,δ condition, the subsequence $\langle a_{f(n)} \rangle$ converges to L.

(ii) The proof of (ii) is similar to (i). ⊣

A sequence $\langle a_n \rangle$ is <u>bounded</u> if its range $\{a_n : n \in N\}$ is a bounded set. Thus by Theorem 5 in Section 1D, $\langle a_n \rangle$ is bounded if and only if every term a_H, $H \in N^*$, is finite.

BOLZANO-WEIERSTRASS THEOREM

Every bounded sequence has a convergent subsequence.

Proof Let $\langle a_n \rangle$ be bounded and let H be positive infinite. Then a_H is finite and by the preceding theorem, $\langle a_n \rangle$ has a subsequence converging to $st(a_H)$. ⊣

COUNTABLE HEINE-BOREL THEOREM

Let $X_0 \supseteq X_1 \supseteq \ldots \supseteq X_n \supseteq \ldots$ be a decreasing chain of nonempty closed bounded sets of real numbers. Then $\bigcap_{n=0}^{\infty} X_n \neq \emptyset$.

Proof Let $P(x,n)$ be the binary relation $x \in X_n$ on the reals. Then for each integer n, $X_n^* = \{x: P^*(x,n)\}$. Choose an infinite H and define $X_H^* = \{x: P^*(x,H)\}$. Each X_n is nonempty, so by the Partial Solution Theorem X_H^* is nonempty and has an element $x \in X_H^*$. Using the Solution Axiom we see that $X_H^* \subseteq X_n^*$ for all $n \leq H$ and hence for all $n \in N$. Therefore $x \in X_n^*$ for all integers n. Since X_n is closed and bounded, x is finite and $st(x) \in X_n$. Thus $st(x) \in \bigcap_{n=0}^{\infty} X_n$. ⊣

THEOREM 6

An increasing sequence $\langle a_n \rangle$ either converges or diverges to ∞.

Proof Suppose the sequence does not diverge to ∞. Then some a_H is not positive infinite. Since $a_H \geq a_0$, a_H is finite. Hence there is a subsequence $\langle a_{f(n)} \rangle$ converging to $L = st(a_H)$. For every integer $n \geq f(0)$ there is an m such that $f(m) \leq n < f(m+1)$. Let K be infinite. By the Partial Solution Theorem there is an M such that $f(M) \leq K \leq f(M+1)$. M is positive infinite, so $a_{f(M)} \approx L \approx a_{f(M+1)}$. But since $\langle a_n \rangle$ is increasing, $a_{f(M)} \leq a_K \leq a_{f(M+1)}$. Therefore $a_K \approx L$. ⊣

9B SERIES (§9.2 - §9.6)

Given an infinite sequence

$$\langle a_n \rangle = a_1, a_2, \ldots, a_n, \ldots,$$

the partial sum sequence $\langle S_n \rangle$ is defined by

$$S_n = a_1 + \ldots + a_n = \sum_{k=1}^{n} a_k .$$

By the Function Axiom, for each positive infinite hyperinteger H we have the <u>infinite partial sum</u>

$$S_H = a_1 + \ldots + a_H = \sum_{k=1}^{H} a_k .$$

DEFINITION

The <u>infinite series</u> $\Sigma_{n=1}^{\infty} a_n$ is said to <u>converge</u> to S, in symbols

$$S = a_1 + \ldots + a_n + \ldots = \sum_{n=1}^{\infty} a_n ,$$

if the partial sum sequence converges to S,

$$S = \lim_{n\to\infty} (a_1 + \ldots + a_n) .$$

That is, for every positive infinite hyperinteger H,

$$S \approx a_1 + \ldots + a_H .$$

The series is said to <u>diverge</u> if the partial sum sequence diverges.

The equivalence theorems for limits of sequences in the preceding section automatically give equivalence theorems for sums of series. In particular, the Hyperreal Cauchy Condition has the following consequence, where $a_H + \ldots + a_K$ means $\Sigma_{n=1}^{K} a_n - \Sigma_{n=1}^{H} a_n$.

THEOREM 1

(i) $\Sigma_{n=1}^{\infty} a_n$ converges if and only if for all infinite $H \le K$, $a_H + \ldots + a_K \approx 0$.

(ii) If $\Sigma_{n=1}^{\infty} a_n$ converges then $\lim_{n\to\infty} a_n = 0$.

<u>Proof</u> (i) is the Hyperreal Cauchy Condition. (ii) follows by setting $K = H$, whence $a_H \approx 0$. ⊣

The usual convergence tests for infinite series are developed in Elementary Calculus in the standard way. The following hyperreal form of the Limit Comparison Test is simpler to state and use than the standard result.

LIMIT COMPARISON TEST

Let $\Sigma_{n=1}^{\infty} a_n$ and $\Sigma_{n=1}^{\infty} b_n$ be positive term series and c a positive real number. Suppose that $a_K \le c\, b_K$ for all infinite K. If $\Sigma_{n=1}^{\infty} b_n$ converges then $\Sigma_{n=1}^{\infty} a_n$ converges.

Proof Let $H \le K$ be infinite. By the Cauchy Condition,

$$b_H + \ldots + b_K \approx 0 .$$

Hence

$$0 \le a_H + \ldots + a_K \le c(b_H + \ldots + b_K) \approx 0 ,$$

so $a_H + \ldots + a_K \approx 0$ and $\Sigma_{n=1}^{\infty} a_n$ converges. ⊣

As an illustration of the use of the Limit Comparison Test we give a proof of a basic result on power series. A series $\Sigma_{n=0}^{\infty} a_n$ is absolutely convergent if the series $\Sigma_{n=0}^{\infty} |a_n|$ is convergent.

THEOREM 2

If a power series $\Sigma_{n=0}^{\infty} a_n x^n$ is convergent when $x = u$, then it is absolutely convergent whenever $|x| < |u|$.

Proof Let $|v| < |u|$ and $b = |v|/|u|$. Then $0 \le b < 1$, so the geometric series $\Sigma_{n=0}^{\infty} b^n$ converges. For positive infinite H, $a_H u^H \approx 0$ since $\Sigma_{n=0}^{\infty} a_n u^n$ converges. Thus

$$|a_H v^H| = |a_H u^H| b^H \le b^H ,$$

so by the Limit Comparison Test, $\Sigma_{n=0}^{\infty} |a_n v^n|$ converges. ⊣

9C TAYLOR'S FORMULA AND HIGHER DIFFERENTIALS (§9.10)

In Elementary Calculus the standard proof of Taylor's Formula is given. Let f be a real function and $c, x \in R$.

TAYLOR'S FORMULA

If $f^{(n+1)}$ exists between c and x, then

$$f(x) = \sum_{k=0}^{n} \frac{f^{(k)}(c)}{k!}(x-c)^k + \frac{f^{(n+1)}(t)}{(n+1)!}(x-c)^{n+1}$$

for some real t between c and x.

Using the Partial Solution Theorem we obtain the following consequence concerning infinitesimals.

COROLLARY 1 Suppose c is real and $f^{(n)}$ is continuous at c. Then whenever $x \approx c$ and Δx is a nonzero infinitesimal,

$$f(x+\Delta x) \approx \sum_{k=0}^{n} \frac{f^{(k)}(x)}{k!}\Delta x^k \quad (\text{compared to } \Delta x^n) .$$

Proof $f^{(n)}(t)$ is defined for $t \approx c$. By the Partial Solution Theorem there is a t between x and $x + \Delta x$ such that

$$f(x+\Delta x) = \sum_{k=0}^{n-1} \frac{f^{(k)}(x)}{k!}\Delta x^k + \frac{f^{(n)}(t)}{n!}\Delta x^n$$

$$= \sum_{k=0}^{n} \frac{f^{(k)}(x)}{k!}\Delta x^k + \frac{f^{(n)}(t) - f^{(n)}(x)}{n!}\Delta x^n .$$

By continuity of $f^{(n)}$ at c, $f^{(n)}(t) - f^{(n)}(x) \approx 0$, and the result follows. ⊣

By modifying the proof of Taylor's Formula, we shall show that Corollary 1 holds at $x = c$ assuming only that $f^{(n)}(c)$ exists.

THEOREM 2

Suppose c is real, Δx is nonzero infinitesimal, and $f^{(n)}(c)$ exists. Then

$$f(c + \Delta x) \approx \sum_{k=0}^{n} \frac{f^{(k)}(c)}{k!} \Delta x^k \quad \text{(compared to } \Delta x^n\text{)} .$$

<u>Proof</u> Since $f^{(n)}(c)$ exists, $f^{(n-1)}(x)$ exists for all x in some real neighborhood of c. Let

$$F(x) = f(x) - \sum_{k=0}^{n} \frac{f^{(k)}(c)}{k!}(x - c)^k ,$$

$$G(x) = (x - c)^n .$$

Then for $m < n$, $F^{(m)}(c) = 0$ and $G^{(m)}(c) = 0$. Using the Generalized Mean Value Theorem (Section 5B) $n - 1$ times we see that

$$\frac{F(x)}{G(x)} = \frac{F^{(n-1)}(t)}{G^{(n-1)}(t)}$$

for some t between c and x. We have

$$\frac{F^{(n-1)}(t)}{G^{(n-1)}(t)} = \frac{f^{(n-1)}(t) - f^{(n-1)}(c)}{n!(t - c)} - \frac{f^{(n)}(c)}{n!} .$$

Therefore

$$f(x) = \sum_{k=0}^{n-1} \frac{f^{(k)}(c)}{k!}(x - c)^k + \frac{f^{(n-1)}(t) - f^{(n-1)}(c)}{n!(t - c)}(x - c)^n .$$

By the Partial Solution Theorem there is a hyperreal t_1 between c and $c + \Delta x$ such that

$$f(c + \Delta x) = \sum_{k=0}^{n-1} \frac{f^{(k)}(c)}{k!} \Delta x^k + \frac{f^{(n-1)}(t_1) - f^{(n-1)}(c)}{n!(t_1 - c)} \Delta x^n .$$

Since $f^{(n)}(c)$ exists, we have

$$f^{(n)}(c) \approx \frac{f^{(n-1)}(t_1) - f^{(n-1)}(c)}{t_1 - c},$$

and the desired result follows. ⊣

When n = 1, Theorem 2 reduces to the Increment Theorem,

$$f(c + \Delta x) \approx f(c) + f'(c)\Delta x \quad \text{(compared to } \Delta x\text{)} .$$

Another natural way to generalize the Increment Theorem involves the notion of an nth increment.

DEFINITION

Let $y = f(x)$ be defined on an interval I. The $n^{\underline{th}}$ increment

$$\Delta^n y = \Delta^n f(x,\Delta x)$$

is defined by induction on n as follows:

$$\Delta' y = \Delta' f(x,\Delta x) = f(x + \Delta x) - f(x) ,$$

$$\Delta^{n+1} y = \Delta^{n+1} f(x,\Delta x) = \Delta^n f(x + \Delta x,\Delta x) - \Delta^n f(x,\Delta x) .$$

Thus $\Delta^n f(x,\Delta x)$ is defined whenever $x \in I$ and $x + n\Delta x \in I$. For n = 2 we have

$$\begin{aligned}\Delta^2 y &= [f(x + 2\Delta x) - f(x + \Delta x)] - [f(x + \Delta x) - f(x)] \\ &= f(x + 2\Delta x) - 2f(x + \Delta x) + f(x).\end{aligned}$$

For n = 3,

$$\Delta^3 y = f(x + 3\Delta x) - 3f(x + 2\Delta x) + 3f(x + \Delta x) - f(x) .$$

In general, it can be shown by induction that

$$\Delta^n y = \sum_{k=0}^{n} (-1)^{n-k} \binom{n}{k} f(x + k\Delta x)$$

where $\binom{n}{k} = \frac{n!}{k!(n-k)!}$ is the binomial coefficient. The following theorem gives a connection between the nth increment and the nth differential $d^n y = f^{(n)}(x)dx^n$.

THEOREM 3

Suppose $y = f(x)$ is a real function, c is a real number, and Δx is nonzero infinitesimal.

(i) If $f^{(n)}$ is continuous at c, then at every hyperreal point $x \approx c$ we have

$$\frac{\Delta^n y}{\Delta x^n} \approx \frac{d^n y}{dx^n},$$

that is,

$$\Delta^n f(x,\Delta x) \approx f^{(n)}(x)\Delta x^n \quad (\text{compared to } \Delta x^n).$$

(ii) If $f^{(n)}$ exists at c, then at $x = c$ we have

$$\frac{\Delta^n y}{\Delta x^n} \approx \frac{d^n y}{dx^n},$$

that is,

$$\Delta^n f(c,\Delta x) \approx f^{(n)}(c)\Delta x^n \quad (\text{compared to } \Delta x^n) .$$

<u>Proof</u> We give the proof for the case $\Delta x > 0$. We first prove a generalization of the Mean Value Theorem.

Let u and Δu be real and suppose $g^{(m)}(t)$ exists for $u \le t \le u + m\Delta u$. Then $\Delta^m g(u,\Delta u) = g^{(m)}(t_1)\Delta u^m$ for some $t_1 \in (u, u + m\Delta u)$. (*)

The proof is by induction on m. For $m = 1$ it is just the Mean Value Theorem,

$$g(u + \Delta u) - g(u) = g'(t)\Delta u.$$

Assume (*) for $m - 1$ and let $h(t) = g(t + \Delta u) - g(t)$. Then

$$h^{(m-1)}(t) = g^{(m-1)}(t + \Delta u) - g^{(m-1)}(t)$$

exists for $u \le t \le u + (m-1)\Delta u$, so for some $t_0 \in (u, u + (m-1)\Delta u)$,

$$\Delta^{m-1}h(u,\Delta u) = h^{(m-1)}(t_0)\Delta u^{m-1} .$$

Moreover,

$$\Delta^{m-1}h(u,\Delta u) = \Delta^m g(u,\Delta u) .$$

By the Mean Value Theorem,

$$h^{(m-1)}(t_0) = g^{(m-1)}(t_0 + \Delta u) - g^{(m-1)}(t_0) = g^{(m)}(t_1)\Delta u$$

for some $t_1 \in (t_0, t_0 + \Delta u)$ and

$$\Delta^m g(u,\Delta u) = g^{(m)}(t_1)\Delta u .$$

This proves (*).

We now prove (i). Since $f^{(n)}$ is continuous at c, $f^{(n)}(t)$ exists for $x \le t \le x + \Delta x$. By (*) and the Partial Solution Theorem,

$$\frac{\Delta^n f(x,\Delta x)}{\Delta x^n} = f^{(m)}(t_1)$$

for some $t_1 \in (x, x + n\Delta x)$. Then $t_1 \approx x \approx c$ and since $f^{(n)}$ is continuous at c,

$$\frac{\Delta^n f(x,\Delta x)}{\Delta x^n} \approx f^{(n)}(x) .$$

To prove (ii) we let $m = n - 1$ and $g(t) = f(t + \Delta x) - f(t)$. By (*) and the Partial Solution Theorem there is a $t_1 \in (c, c + (n-1)\Delta x)$ such that

$$\Delta^{n-1}g(c,\Delta x) = g^{(n-1)}(t_1)\Delta x^{n-1} ,$$

that is,

$$\Delta^n f(c,\Delta x) = \frac{f^{(n-1)}(t_1 + \Delta x) - f^{(n-1)}(t_1)}{\Delta x}\Delta x^n .$$

Let $t_2 = t_1 + \Delta x$. Then $t_1 \approx c$, $t_2 \approx c$, so

$$\frac{\Delta^n f(c,\Delta x)}{\Delta x^n} = \frac{f^{(n-1)}(t_2) - f^{(n-1)}(t_1)}{\Delta x}$$

$$= \frac{f^{(n-1)}(t_2) - f^{(n-1)}(c)}{t_2 - c}\,\frac{t_2 - c}{\Delta x}$$

$$+ \frac{f^{(n-1)}(c) - f^{(n-1)}(t_1)}{c - t_1}\,\frac{c - t_1}{\Delta x}$$

$$\approx f^{(n)}(c)\,\mathrm{st}\left(\frac{t_2 - c}{\Delta x}\right) + f^{(n)}(c)\,\mathrm{st}\left(\frac{c - t_1}{\Delta x}\right) = f^{(n)}(c) \;.$$

Thus

$$\frac{\Delta^n f(c,\Delta x)}{\Delta x^n} \approx f^{(n)}(c) \;.$$

10 VECTORS

10A HYPERREAL VECTORS

Chapter 10 in <u>Elementary Calculus</u> deals mostly with two and three dimensional vectors over the reals. Only the last section of the chapter, §10.8, concerns hyperreal vectors.

Let n be a fixed positive integer. The simplest way to define a real vector in n dimensions is as an n-tuple of real numbers, $\langle r_1,\ldots,r_n\rangle$. In <u>Elementary Calculus</u> we used the following more geometric definition. Given two points

$$P = \langle p_1,\ldots,p_n\rangle,\ Q = \langle q_1,\ldots,q_n\rangle$$

in R^n, the ordered pair $\overrightarrow{PQ}$ is called the <u>directed line segment</u> from P to Q. The <u>components</u> of $\overrightarrow{PQ}$ are the terms in the n-tuple

$$\langle q_1 - p_1,\ldots,q_n - p_n\rangle.$$

Two directed line segments are said to be <u>equivalent</u> if they have the same components (hence the same length and direction), and a <u>vector</u> $\vec{A}$ is an equivalence class of directed line segments. If $\vec{A}$ is the equivalence class of $\overrightarrow{PQ}$, the components of $\overrightarrow{PQ}$ are called the <u>components of</u> $\vec{A}$, denoted by $\langle a_1,\ldots,a_n\rangle$, and $\vec{A}$ is called the <u>vector from</u> P <u>to</u> Q.

The <u>basis vectors</u> are the vectors $\vec{j}_1,\ldots,\vec{j}_n$, where $\vec{j}_m$ is the vector with m^{th} component 1 and all other components 0. <u>Vector sums</u> $\vec{A} + \vec{B}$ and <u>scalar multiples</u> $c\vec{A}$ of real vectors can be defined in the usual way in terms of the components. Thus the vector $\vec{A}$ with components $\langle a_1,\ldots,a_n\rangle$ is equal to the sum

$$\vec{A} = a_1\vec{j}_1 + \ldots + a_n\vec{j}_n .$$

The inner product and length are defined by

$$\vec{A}\cdot\vec{B} = a_1b_1 + \dots + a_nb_n ,$$

$$|\vec{A}| = \sqrt{\vec{A}\cdot\vec{A}} = \sqrt{a_1^2 + \dots + a_n^2} .$$

The zero vector in n dimensions is denoted by $\vec{0}$.

Hyperreal vectors in n dimensions are defined in a similar way. By the Function and Solution Axioms, the hyperreal vectors in n dimensions form an inner product space over the field R* of hyperreal numbers. That is, the usual algebraic rules for vector sums, scalar multiples, and inner products hold for hyperreal vectors. Hyperreal vectors $\vec{A}$ will be classified by the behavior of the length $|\vec{A}|$ and the unit vector $\frac{1}{|\vec{A}|}\vec{A}$.

DEFINITION

A hyperreal vector $\vec{A}$ is said to be infinitesimal, finite, or infinite if its length $|\vec{A}|$ is infinitesimal, finite, or infinite respectively. $\vec{A}$ is infinitely close to $\vec{B}$, $\vec{A} \approx \vec{B}$, if and only if $\vec{B} - \vec{A}$ is infinitesimal. The monad and galaxy of $\vec{A}$ are defined by

$$\text{monad}(\vec{A}) = \{\vec{B} : \vec{B} \approx \vec{A}\},$$

$$\text{galaxy}(\vec{A}) = \{\vec{B} : \vec{B} - \vec{A} \text{ is finite}\}.$$

Thus monad($\vec{0}$) is the set of infinitesimal vectors and galaxy($\vec{0}$) the set of finite vectors.

PROPOSITION 1

(i) $\vec{A}$ is infinitesimal if and only if all its components are infinitesimal.

(ii) $\vec{A}$ is finite if and only if all its components are finite.

(iii) $\vec{A}$ is infinite if and only if at least one of its components is infinite.

(iv) $\vec{A} \approx \vec{B}$ if and only if $a_1 \approx b_1, \dots, a_n \approx b_n$.

The above result follows easily from the inequalities

$$|a_m| \le \sqrt{a_1^2 + \dots + a_m^2} \le |a_1| + \dots + |a_m|.$$

PROPOSITION 2

The sets monad($\vec{0}$) and galaxy($\vec{0}$) are closed under vector sums and finite scalar multiples.

This follows from Proposition 1 and the fact that the sets monad(0) and galaxy(0) of hyperreal numbers are closed under sums and finite scalar multiples.

DEFINITION

Given a finite hyperreal vector $\vec{A}$, the <u>standard part</u> $st(\vec{A})$ is defined as the real vector

$$st(\vec{A}) = st(a_1)j_1 + \dots + st(a_n)j_n \ .$$

Thus $st(\vec{A})$ is the unique real vector infinitely close to $\vec{A}$.

The next result also follows from Proposition 1.

PROPOSITION 3

The mapping st preserves sums, finite scalar multiples, inner products, and lengths. That is, if $\vec{A}$, $\vec{B}$ are finite vectors and c is a finite scalar,

$$st(\vec{A} + \vec{B}) = st(\vec{A}) + st(\vec{B})$$

$$st(c\vec{A}) = st(c)\,st(\vec{A})$$

$$st(\vec{A} \cdot \vec{B}) = st(\vec{A}) \cdot st(\vec{B})$$

$$st(|\vec{A}|) = |st(\vec{A})| \ .$$

DEFINITION

A hyperreal vector $\vec{A}$ has <u>real length</u> if $|\vec{A}|$ is real. $\vec{A}$ has <u>real direction</u> if $\vec{A} \neq \vec{0}$ and the <u>unit vector</u> $\vec{U} = \frac{1}{|\vec{A}|}\vec{A}$ is real.

PROPOSITION 4

A nonzero vector $\vec{A}$ is real if and only if $\vec{A}$ has both real length and real direction.

Proof $\vec{A}$ has real length and direction if and only if both $\vec{A}$ and $\vec{U} = \frac{1}{|\vec{A}|}\vec{A}$ are real if and only if $\vec{A} = |\vec{A}|\vec{U}$ is real.

Two nonzero vectors $\vec{A}$ and $\vec{B}$ are said to be parallel, $\vec{A}\,||\,\vec{B}$, if their unit vectors $\vec{U}$ and $\vec{V}$ are equal or opposite, $\vec{U} = \vec{V}$ or $\vec{U} = -\vec{V}$.

We introduce a weaker notion for hyperreal vectors. Two nonzero hyperreal vectors $\vec{A}$ and $\vec{B}$ are said to be almost parallel if their unit vectors $\vec{U}$ and $\vec{V}$ are such that $\vec{U} \approx \vec{V}$ or $\vec{U} \approx -\vec{V}$.

PROPOSITION 5

Every nonzero hyperreal vector $\vec{A}$ is almost parallel to some real vector.

Proof Let $\vec{U}$ be the unit vector of $\vec{A}$. Then $\vec{U}$ has finite length 1, so the real vector $\vec{V} = st(\vec{U})$ exists. $\vec{V}$ is its own unit vector because $|\vec{V}| = |st(\vec{U})| = st(|\vec{U}|) = 1$. Finally, $\vec{U} \approx \vec{V}$, so $\vec{A}$ is almost parallel to $\vec{V}$. ⊣

The notion of almost parallel can be generalized to finite sequences of vectors as follows. A k-tuple of real vectors $\vec{A}_1, \ldots, \vec{A}_k$ is linearly dependent over R if there exist real numbers $c_1, \ldots, c_k$, not all zero, such that

$$c_1\vec{A}_1 + \ldots + c_k\vec{A}_k = \vec{0} . \tag{1}$$

Similarly a k-tuple $\vec{A}_1, \ldots, \vec{A}_k$ of hyperreal vectors is linearly dependent over R* if there exist hyperreal numbers $c_1, \ldots, c_k$, not all zero, such that (1) holds. Finally, a k-tuple of hyperreal vectors $\vec{A}_1, \ldots, \vec{A}_k$ is almost linearly dependent over R if either some $\vec{A}_m = \vec{0}$ or there exist real numbers $c_1, \ldots, c_k$, not all zero, such that

$$c_1\vec{U}_1 + \ldots + c_k\vec{U}_k \approx \vec{0} ,$$

where $\vec{U}_m$ is the unit vector of $\vec{A}_m$.

Thus nonzero vectors $\vec{A}_1,\ldots,\vec{A}_k$ are almost linearly dependent over R if and only if the standard parts of the unit vectors, $st(\vec{U}_1),\ldots,st(\vec{U}_n)$, are linearly dependent over R. For pairs of nonzero vectors, linearly dependent means parallel and almost linearly dependent means almost parallel.

THEOREM 6

(i) If $\vec{A}_1,\ldots,\vec{A}_k$ are linearly dependent over R* then they are almost linearly dependent over R.

(ii) For real vectors, linear dependence over R, linear dependence over R*, and almost linear dependence over R are equivalent.

Proof (i) Let $c_1\vec{A}_1 + \ldots + c_k\vec{A}_k = \vec{0}$ where the c_k are not all zero. We may assume all the $\vec{A}_m \neq \vec{0}$ since otherwise the vectors are trivially almost linearly dependent over R. Then

$$c_1|\vec{A}_1|\vec{U}_1 + \ldots + c_k|\vec{A}_k|\vec{U}_k = \vec{0}\ .$$

Let $b = |c_m||\vec{A}_m|$ be the largest of the k numbers $|c_1||\vec{A}_1|,\ldots,$ $|c_k||\vec{A}_k|$, and let $d_\ell = c_\ell|\vec{A}_\ell|/b$ for $\ell = 1,\ldots,k$. Then

$$d_1\vec{U}_1 + \ldots + d_k\vec{U}_k = \vec{0}\ ,$$

each d_ℓ is finite, and $|d_m| = 1$. Taking standard parts we see that

$$st(d_1)st(\vec{U}_1) + \ldots + st(d_k)st(\vec{U}_k) = \vec{0}$$

and $st(d_m) \neq 0$. Therefore the standard parts of the unit vectors are linearly dependent over R, hence $\vec{A}_1,\ldots,\vec{A}_k$ are almost linearly dependent over R. This proves (i).

By the Partial Solution Theorem, for real vectors, linear dependence over R and R* are equivalent. Moreover, real vectors are linearly dependent over R if and only if the standard parts of their unit vectors are linearly dependent over R. This proves (ii). ⊣

10B VECTOR FUNCTIONS

A <u>real vector function</u> $\vec{F}$ maps a subset of R into the set of n dimensional real vectors. The components of a real vector function $\vec{F}$ are real functions $<f_1,\ldots,f_n>$ with the same domain as $\vec{F}$. In symbols, $\vec{F}(t) = f_1(t)\vec{j}_1 + \ldots + f_n(t)\vec{j}_n$. The natural extension of $\vec{F}$ is the hyperreal vector function $\vec{F}^*$ whose components are the natural extensions $\langle f_1^*, \ldots, f_n^* \rangle$.

We conclude this chapter with definitions and equivalence theorems for limits, continuity, and derivatives of real vector functions of one variable.

DEFINITION

Let $\vec{F}$ be a real vector function, $\vec{A}$ a real vector, and c a real number.

$$\lim_{t \to c} \vec{F}(t) = \vec{A}$$

means that whenever $t \approx c$ but $t \neq c$, $\vec{F}(t) \approx \vec{A}$. $\vec{F}$ is <u>continuous</u> at c if whenever $t \approx c$, $\vec{F}(t) \approx \vec{F}(c)$, i.e.,

$$\lim_{t \to c} \vec{F}(t) = \vec{F}(c).$$

THEOREM 7

The following are equivalent.

(i) $\lim_{t \to c} \vec{F}(t) = \vec{A}$.

(ii) For $m = 1, \ldots, n$, $\lim_{t \to c} f_m(t) = a_m$.

(iii) The ε, δ condition. For every real $\varepsilon > 0$ there is a real $\delta > 0$ such that whenever $0 < |t - c| < \delta$, $|\vec{F}(t) - \vec{A}| < \varepsilon$.

<u>Proof</u> The equivalence of (i) and (ii) follows from Proposition 1, and the equivalence of (ii) and (iii) follows from the equivalence theorem for limits of real functions. ⊣

DEFINITION

Let $\vec{F}$ be a real function, $\vec{S}$ a real vector, and c a real number. $\vec{F}$ has <u>vector derivative</u> $\vec{S}$ at c, in symbols $\vec{F}'(c) = \vec{S}$, if for every nonzero infinitesimal Δt we have

$$\frac{\vec{F}(c + \Delta t) - \vec{F}(c)}{\Delta t} \approx \vec{S} .$$

Thus

$$\vec{F}'(c) = \lim_{\Delta t \to 0} \frac{\vec{F}(c + \Delta t) - \vec{F}(c)}{\Delta t} .$$

The Equivalence Theorem for Limits leads to equivalent conditions for the vector derivative. In particular it follows that $\vec{F}'(c)$ exists if and only if $f_1'(c),\ldots,f_n'(c)$ all exist, and

$$\vec{F}'(c) = f_1'(c)\vec{j}_1 + \ldots + f_n'(c)\vec{j}_n .$$

Vector increments and vector differentials are defined as vector dependent variables as follows. We are given a real vector function $\vec{X} = \vec{F}(t)$ where t is a scalar independent variable and $\vec{X}$ a vector dependent variable. We introduce a new scalar independent variable Δt and a new vector dependent variable $\Delta\vec{X}$, called the vector increment of $\vec{X}$, with the equation

$$\Delta\vec{X} = \vec{F}(t + \Delta t) - \vec{F}(t) .$$

The vector differential of $\vec{X}$ is a second vector dependent variable whose values are given by the equation

$$d\vec{X} = \vec{F}'(t)\Delta t .$$

Thus $d\vec{X}$ exists when $\vec{F}'(t)$ exists, and putting $dt = \Delta t$ we have

$$\frac{d\vec{X}}{dt} = \vec{F}'(t), \quad \frac{\Delta\vec{X}}{\Delta t} \approx \frac{d\vec{X}}{dt} .$$

11 PARTIAL DIFFERENTIATION

11A CONTINUITY AND PARTIAL DERIVATIVES (§11.1 - §11.4)

To simplify notation we concentrate on real functions of two variables. However, all the notions and results readily extend to n variables.

Recall from Section 1D that the distance between two hyperreal points $P(x_1,y_1)$ and $Q(x_2,y_2)$ is

$$|P - Q| = \sqrt{(x_2 - x_1)^2 + (y_2 - y_1)^2} ,$$

and that P is infinitely close to Q, $P \approx Q$, if $|P - Q| \approx 0$. The monad of P is the set of all points Q infinitely close to P. We have $P(x_1,y_1) \approx Q(x_2,y_2)$ if and only if $x_1 \approx x_2$ and $y_1 \approx y_2$.

Given a real point P and a real number $\delta > 0$, the real neighborhood $N_\delta(P)$ is defined as the set of all real points Q such that $|Q - P| < \delta$. The interior of a set $D \subseteq R^2$ is the set of all points P such that some real neighborhood $N_\delta(P)$ is contained in D. A set D is open if it is equal to its interior. A point P belongs to the boundary of D if every real neighborhood of P meets both D and its complement. A set D is closed if D contains its boundary.

The following result was proved for one variable in Section 1D, and the proof for two variables is similar.

THEOREM 1

(i) Let $Y \subseteq R^2$ be a set of real points and let $P \in R^2$ be a real point. If Y* contains the monad of P, then Y contains some real neighborhood of P.

(ii) Let f be a real function of two variables and $P \in R^2$. If $f(x,y)$ is defined for all $(x,y) \approx P$, then f is defined on some real neighborhood of P.

Hereafter let f be a real function of two variables.

DEFINITION

Let $(a,b) \in R^2$ and $L \in R$.

$$\lim_{(x,y)\to(a,b)} f(x,y) = L$$

means that whenever $(x,y) \approx (a,b)$ but $(x,y) \neq (a,b)$, $f(x,y) \approx L$. Infinite limits are defined in a similar way.

f is <u>continuous</u> at (a,b) if f(a,b) exists and

$$\lim_{(x,y)\to(a,b)} f(x,y) = f(a,b).$$

It follows from Theorem 1 that any function which is continuous at (a,b) is defined on some real neighborhood of (a,b). As in the case of one variable, we have the following equivalence theorem.

THEOREM 2

The following are equivalent.

(i) $\lim_{(x,y)\to(a,b)} f(x,y) = L.$

(ii) For every real $\varepsilon > 0$ there is a real $\delta > 0$ such that for all $(x,y) \neq (a,b)$ in $N_\delta(a,b)$, $|f(x,y) - L| < \varepsilon$.

(iii) There is a hyperreal $\delta > 0$ such that whenever

$$0 < |(x,y) - (a,b)| < \delta,$$

we have $f(x,y) \approx L$.

The result that st is a ring homomorphism from galaxy(0) onto R shows that the functions x + y, x − y, and xy are everywhere continuous. For example, if $(x,y) \approx (a,b) \in R^2$ then

$$x + y \approx st(x + y) = st(x) + st(y) = a + b\ .$$

THEOREM 3

Compositions of continuous functions are continuous. That is, if $f(x,y)$ and $g(x,y)$ are continuous at $(x,y) = (a,b)$, and if $h(u,v)$ is continuous at $(u,v) = (f(a,b),g(a,b))$, then the composition

$$k(x,y) = h(f(x,y),g(x,y))$$

is continuous at $(x,y) = (a,b)$.

<u>Proof</u> Let $(x,y) \approx (a,b)$. Then $f(x,y) \approx f(a,b)$, $g(x,y) \approx g(a,b)$, hence

$$k(x,y) = h(f(x,y),g(x,y)) \approx h(f(a,b),g(a,b)) = k(a,b)\ . \dashv$$

DEFINITION

Given a real function $f(x,y)$ and a real point (a,b), the <u>partial derivatives</u> are defined by

$$f_x(a,b) = g'(a) \text{ where } g(x) = f(x,b),$$

$$f_y(a,b) = h'(b) \text{ where } h(y) = f(a,y)\ .$$

If $z = f(x,y)$ we use the notation

$$\frac{\partial z}{\partial x} = \frac{\partial f}{\partial x}(a,b) = f_x(a,b),\ \frac{\partial z}{\partial y} = \frac{\partial f}{\partial y}(a,b) = f_y(a,b)\ .$$

The mere existence of the partial derivatives of $f(x,y)$ at (a,b) tells us nothing about the behavior of f off of the lines $x = a$, $y = b$. We introduce three stronger notions of differentiability which will be used later on in this chapter.

DEFINITION

Let $(a,b) \in R^2$ and suppose both partial derivatives $f_x(a,b)$ and $f_y(a,b)$ exist.

(i) f is <u>smooth</u> at (a,b) if both f_x and f_y are continuous at (a,b).

(ii) f is <u>uniformly differentiable</u> at (a,b) if for any point $(x,y) \approx (a,b)$ and nonzero infinitesimal $(\Delta x,\Delta y) \underset{\neq}{\approx} (0,0)$,

$$f(x + \Delta x,y + \Delta y) - f(x,y) \approx f_x(a,b)\Delta x + f_y(a,b)\Delta y$$

$$\text{(compared to } \sqrt{\Delta x^2 + \Delta y^2}\text{)}.$$

(iii) f is <u>differentiable</u> at (a,b) if for whenever $(\Delta x,\Delta y) \underset{\neq}{\approx} (0,0)$,

$$f(a + \Delta x,b + \Delta y) - f(a,b) \approx f_x(a,b)\Delta x + f_y(a,b)\Delta y$$

$$\text{(compared to } \sqrt{\Delta x^2 + \Delta y^2}\text{)}.$$

Uniform differentiability and differentiability correspond to the one variable notions. They are equivalent to real ε,δ conditions which are left to the reader.

When $z = f(x,y)$ we introduce two new dependent variables, the <u>increment</u> Δz given by

$$\Delta z = f(x + \Delta x,y + \Delta y) - f(x,y)$$

and the <u>total differential</u> dz given by

$$dz = f_x(x,y)\Delta x + f_y(x,y)\Delta y = \frac{\partial z}{\partial x}\Delta x + \frac{\partial z}{\partial y}\Delta y .$$

Both Δz and dz depend on the four independent variables x, y, Δx, Δy. Using dependent variable notation, f is differentiable at (a,b) if and only if whenever $(\Delta x,\Delta y) \underset{\neq}{\approx} (0,0)$,

$$\Delta z \approx dz \quad \text{(compared to } \sqrt{\Delta x^2 + \Delta y^2}\text{)}.$$

If f is differentiable at (a,b), the plane with the equation

$$z - f(a,b) = f_x(a,b)(x - a) + f_y(a,b)(y - b)$$

is called the tangent plane of f at (a,b). Putting $\Delta x = x - a$ and $\Delta y = y - b$, we see that

$$\Delta z = \text{change in z along the surface,}$$

$$dz = \text{change in z along the tangent plane.}$$

By definition, if f is differentiable at (a,b) then whenever $(x,y) \underset{\neq}{\approx} (a,b)$, the tangent plane is infinitely close to the surface compared to $\sqrt{\Delta x^2 + \Delta y^2}$.

The following result shows the relationship between the three notions of differentiability. The implication (i) → (iii) was called the Increment Theorem in Elementary Calculus.

THEOREM 4

Let (a,b) be a real point. Each condition below implies the next, (i) → (ii) → (iii) → (iv).

(i) f is smooth at (a,b).
(ii) f is uniformly differentiable at (a,b).
(iii) f is differentiable at (a,b).
(iv) f is continuous at (a,b).

Proof (i) → (ii). Let $(x,y) \approx (a,b)$ and let $(\Delta x,\Delta y) \underset{\neq}{\approx} (0,0)$. Then

$$f(x + \Delta x,y + \Delta y) - f(x,y) = [f(x + \Delta x,y + \Delta y) - f(x + \Delta x,y)] + [f(x + \Delta x,y) - f(x,y)].$$

Since f_x and f_y are continuous at (a,b) they are defined everywhere in the monad of (a,b). Using the Hyperreal Mean Value Theorem for the one variable function $g(u) = f(x + \Delta x,u)$, we have

$$f(x + \Delta x,y + \Delta y) - f(x + \Delta x,y) = f_y(x + \Delta x,u)\Delta y$$

for some u between y and $y + \Delta y$. Similarly,

$$f(x + \Delta x,y) - f(x,y) = f_x(t,y)\Delta x$$

for some t between x and $x + \Delta x$. Since

$$f_y(x + \Delta x,u) \approx f_y(a,b), \; f_x(t,y) \approx f_x(a,b)$$

and

$$\frac{\Delta x}{\sqrt{\Delta x^2 + \Delta y^2}}, \quad \frac{\Delta y}{\sqrt{\Delta x^2 + \Delta y^2}}$$

are finite, we have

$$f(x + \Delta x, y + \Delta y) - f(x,y) \approx f_x(a,b)\Delta x + f_y(a,b)\Delta y$$

$$\text{(compared to } \sqrt{\Delta x^2 + \Delta y^2}\text{).}$$

Thus (i) implies (ii). Condition (ii) trivially implies (iii).

(iii) → (iv). Let $(\Delta x, \Delta y) \approx 0$. By (iii),

$$f(a + \Delta x, b + \Delta y) - f(a,b) \approx f_x(a,b)\Delta x + f_y(a,b)\Delta y \approx 0.$$

Therefore f is continuous at (a,b). ⊣

THEOREM 5

If f(x,y) is uniformly differentiable at a real point (a,b), then f is continuous on some real neighborhood of (a,b).

The proof is exactly like the one variable analogue, Theorem 3 in Section 7A.

The following is a two variable form of Theorem 3 in Section 3C on continuous derivatives.

THEOREM 6

Let f(x,y) be a real function whose partial derivatives f_x and f_y exist everywhere in an open set $Y \subseteq R^2$. The following are equivalent.

(i) f is smooth at every point of Y.

(ii) f is uniformly differentiable at every point of Y.

The proof is left to the reader. Note that the direction (i) → (ii) follows from Theorem 3.

In Elementary Calculus the theory was simplified by emphasizing the notion of a smooth function. Some theorems were stated under the hypothesis that f is smooth when only uniform differentiability or differentiability was actually needed. In the next section we give detailed proofs of the basic results and keep track of the kind of differentiability which is actually needed in each case.

11B CHAIN RULE AND IMPLICIT FUNCTIONS (§11.5, §11.6)

The Chain Rule for functions of two variables holds when all functions involved are differentiable in the sense of the preceding section.

CHAIN RULE FOR TWO VARIABLES

Suppose $x = f(s,t)$ and $y = g(s,t)$ are differentiable at the real point (s_0,t_0) and $z = h(x,y)$ is differentiable $(x_0,y_0) = (f(s_0,t_0),g(s_0,t_0))$. Then the composition

$$z = H(s,t) = h(f(s,t),g(s,t))$$

is differentiable at (s_0,t_0), and its partial derivatives are

$$\frac{\partial z}{\partial s} = \frac{\partial z}{\partial x}\frac{\partial x}{\partial s} + \frac{\partial z}{\partial y}\frac{\partial y}{\partial s}, \quad \frac{\partial z}{\partial t} = \frac{\partial z}{\partial x}\frac{\partial x}{\partial t} + \frac{\partial z}{\partial y}\frac{\partial y}{\partial t} .$$

<u>Proof</u> Let $(\Delta s,\Delta t) \underset{\neq}{\approx} (0,0)$ and let Δx, Δy, and Δz be the corresponding increments in x, y, and z. Let $\delta = \sqrt{\Delta s^2 + \Delta t^2}$. Then at the point (s_0,t_0),

$$\Delta x \approx \frac{\partial x}{\partial s}\Delta s + \frac{\partial x}{\partial t}\Delta t \quad (\text{compared to } \delta) ,$$

$$\Delta y \approx \frac{\partial y}{\partial s}\Delta s + \frac{\partial y}{\partial t}\Delta t \quad (\text{compared to } \delta) .$$

It follows that $\Delta x \approx 0$, $\Delta y \approx 0$. Also, $\Delta x/\delta$, $\Delta y/\delta$, and hence $\sqrt{\Delta x^2 + \Delta y^2} \,/\, \delta$ are finite. Therefore

$$\Delta z \approx \frac{\partial z}{\partial x}\Delta x + \frac{\partial z}{\partial y}\Delta y \quad (\text{compared to } \delta) .$$

When $\Delta t = 0$ we have $\delta = \Delta s$ and

$$\frac{\Delta z}{\Delta s} \approx \frac{\partial z}{\partial x}\frac{\Delta x}{\Delta s} + \frac{\partial z}{\partial y}\frac{\Delta y}{\Delta s} \approx \frac{\partial z}{\partial x}\frac{\partial x}{\partial s} + \frac{\partial z}{\partial y}\frac{\partial y}{\partial s},$$

so

$$\frac{\partial z}{\partial s} = \frac{\partial z}{\partial x}\frac{\partial x}{\partial s} + \frac{\partial z}{\partial y}\frac{\partial y}{\partial s}.$$

The formula for $\frac{\partial z}{\partial t}$ is obtained in a similar way when $\Delta s = 0$. Finally, H is differentiable at (s_0,t_0) because

$$\frac{\Delta z}{\delta} \approx \frac{\partial z}{\partial x}\frac{\Delta x}{\delta} + \frac{\partial z}{\partial y}\frac{\Delta y}{\delta}$$

$$\approx \frac{\partial z}{\partial x}\left(\frac{\partial x}{\partial s}\Delta s + \frac{\partial x}{\partial t}\Delta t\right)\frac{1}{\delta} + \frac{\partial z}{\partial y}\left(\frac{\partial y}{\partial s}\Delta s + \frac{\partial y}{\partial t}\Delta t\right)\frac{1}{\delta},$$

so

$$\Delta z \approx \frac{\partial z}{\partial s}\Delta s + \frac{\partial z}{\partial t}\Delta t \quad \text{(compared to } \delta).$$ ⊣

We now turn to the Implicit Function Theorem, first for two variables and then for three variables.

DEFINITION

An implicit function of a real curve $F(x,y) = 0$ at a real point (a,b) is a real function $y = g(x)$ such that:

(i) $g(a) = b$;

(ii) The domain of g is a real neighborhood of a;

(iii) The graph of g is a subset of the graph of the equation $F(x,y) = 0$.

An implicit function of a real surface $F(x,y,z) = 0$ at a real point (a,b,c) is a real function $z = h(x,y)$ such that:

(i') $h(a,b) = c$;

(ii') The domain of h is a real neighborhood of (a,b);

(iii') The graph of h is a subset of the graph of $F(x,y,z) = 0$.

The notion of uniform differentiability is useful in the Implicit Function Theorem.

IMPLICIT FUNCTION THEOREM FOR TWO VARIABLES

Suppose that at the real point (a,b), $z = F(x,y)$ is uniformly differentiable, $F(a,b) = 0$, and $F_y(a,b) \neq 0$. Then the curve $F(x,y) = 0$ has an implicit function at (a,b), every implicit function $g(x)$ is uniformly differentiable at a, and

$$g'(a) = -\frac{F_x(a,b)}{F_y(a,b)} .$$

<u>Proof</u> Since $F_y(a,b) \neq 0$, it follows from the ε,δ condition for uniform differentiability at (a,b) that there is a real neighborhood $N_\delta(a,b)$ with the following property. For any two distinct points (x,y) and $(x,y + \Delta y)$ of $N_\delta(a,b)$, $F(x,y) \neq F(x,y + \Delta y)$. We may therefore define a real function g by

$$g(x) = y \quad \text{iff} \quad (x,y) \in N_\delta(a,b) \text{ and } F(x,y) = 0 .$$

Obviously $g(a) = b$. We show first that g is continuous at a. Let $x_1 \approx a$. Since F is continuous at (a,b), $F(x_1,b) \approx 0$. By uniform differentiability, for every nonzero infinitesimal Δy we have

$$\frac{F(x_1,b + \Delta y) - F(x_1,b)}{\Delta y} \approx F_y(a,b) \neq 0 .$$

Hence there is a $\Delta y \approx 0$ such that 0 is between $F(x_1,b)$ and $F(x_1, b + \Delta y)$. Since F is continuous on $N_\delta(a,b)$, it follows from the Hyperreal Intermediate Value Theorem that there is a point $y_1 \approx b$ with $F(x_1,y_1) = 0$. Every real solution of

$$F(x,y) = 0, \quad |(x,y) - (a,b)| < \delta$$

is a real solution of $y = g(x)$. By the Solution Axiom, $y_1 = g(x_1)$. Thus g is defined in the monad of a and is continuous at a. By Theorem 1 in Section 11A (or Corollary 4 in Section 1D), g is defined on some real neighborhood of a. By definition, the graph of g is a subset of the graph of $F(x,y) = 0$.

It remains to show that g is uniformly differentiable at a and has the required derivative. Let $x \approx a$ and $\Delta x \underset{\neq}{\approx} 0$, and let $y = g(x)$, $\Delta y = g(x + \Delta x) - g(x)$, $\Delta s = \sqrt{\Delta x^2 + \Delta y^2}$. Since g is continuous at a, $\Delta y \approx 0$. By uniform differentiability,

$$\begin{aligned} 0 &= F(x + \Delta x,y + \Delta y) - F(x,y) \\ &\approx F_x(a,b)\Delta x + F_y(a,b)\Delta y \quad \text{(compared to } \Delta s). \end{aligned}$$

Since $F_y(a,b) \neq 0$,

$$\frac{\Delta y}{\Delta s} \approx -\frac{F_x(a,b)}{F_y(a,b)} \frac{\Delta x}{\Delta s}. \tag{1}$$

We cannot have $\Delta x/\Delta s \approx 0$, because it would follow that

$$1 \approx \Delta y/\Delta s \approx -\frac{F_x(a,b)}{F_y(a,b)} \cdot 0 .$$

Therefore we may divide equation (1) by $\Delta x/\Delta s$,

$$\frac{\Delta y}{\Delta x} = \frac{\Delta y/\Delta s}{\Delta x/\Delta s} \approx -\frac{F_x(a,b)}{F_y(a,b)} .$$

We conclude that g is uniformly differentiable at a and $g'(a) = -F_x(a,b)/F_y(a,b)$. ⊣

Before stating the Implicit Function Theorem for three variables we need the notion of uniform differentiability in three variables.

DEFINITION

A real function $F(x,y,z)$ is <u>uniformly differentiable</u> at (a,b,c) if all three partial derivatives of F exist at (a,b,c) and for every hyperreal point $(x,y,z) \approx (a,b,c)$ and nonzero infinitesimal $(\Delta x,\Delta y,\Delta z) \underset{\neq}{\approx} (0,0,0)$,

$$F(x + \Delta x, y + \Delta y, z + \Delta z) - F(x,y,z) \approx F_x(a,b,c)\Delta x + F_y(a,b,c)\Delta y + F_z(a,b,c)\Delta z$$
$$(\text{compared to } \sqrt{\Delta x^2 + \Delta y^2 + \Delta z^2}).$$

The following theorem is proved in exactly the same way as the two variable theorem.

IMPLICIT FUNCTION THEOREM FOR THREE VARIABLES

Suppose that at the point (a,b,c), $w = F(x,y,z)$ is uniformly differentiable, $F(a,b,c) = 0$, and $F_z(a,b,c) \neq 0$. Then the curve

$F(x,y,z) = 0$ has an implicit function at (a,b). Moreover, every implicit function $h(x,y)$ is uniformly differentiable at (a,b), has partial derivatives

$$h_x(a,b) = -\frac{F_x(a,b,c)}{F_z(a,b,c)}, \quad h_y(a,b) = -\frac{F_y(a,b,c)}{F_z(a,b,c)}$$

and has the tangent plane

$$F_x(a,b,c)(x-a) + F_y(a,b,c)(y-b) + F_z(a,b,c)(z-c) = 0.$$

11C MAXIMA AND MINIMA (§11.7)

We discuss maxima and minima for real functions of two variables.

DEFINITION

Let $f(x,y)$ be a real function with domain D. f has a <u>maximum</u> at a point $(x_0,y_0) \in D$ if $f(x_0,y_0) \geq f(x,y)$ for all $(x,y) \in D$. The value $f(x_0,y_0)$ is called the <u>maximum value</u> of f. A <u>minimum</u> of f and the <u>minimum value</u> of f are defined analogously.

The following two results are proved exactly as in the one variable case.

THEOREM 1

Suppose a real function $f(x,y)$ with domain D has a maximum at a real point (a,b). Then $f(a,b) \geq f(x,y)$ for all hyperreal points $(x,y) \in D^*$.

THEOREM 2

Given a real function $f(x,y)$ and a real point (a,b), the following are equivalent.

(i) f has a local maximum at (a,b), that is, the restriction of f to some real neighborhood of (a,b) has a maximum at (a,b).

(ii) Whenever $(x,y) \approx (a,b)$, $f(a,b) \geq f(x,y)$.

Recall that in Section 1D we proved that a set of points D is closed and bounded if and only if:

(i) Every point $(x,y) \in D^*$ is finite;

(ii) Whenever $(x,y) \in D^*$, $(st(x),st(y)) \in D$.

We use this fact to prove the Extreme Value Theorem for two variables.

EXTREME VALUE THEOREM

Suppose the domain of $f(x,y)$ is a closed and bounded set D, and $f(x,y)$ is continuous on D. Then f has a maximum and a minimum.

Proof Let Y be the range of f,

$$Y = \{f(x,y) : (x,y) \in D\}.$$

By Theorem 1 in Section 1D, f^* has domain D^* and range Y^*. For each $(x_1,y_1) \in D^*$, we have $(st(x_1),st(y_1)) \in D$, and by continuity of f,

$$f(x_1,y_1) \approx f(st(x_1),st(y_1)) .$$

Therefore $f(x_1,y_1)$ is finite, so every element of Y^* is finite. By Theorem 5 in Section 1D, the set Y is bounded, say $Y \subseteq [A,B]$ for some real A and B. Consider a positive integer n and partition [A,B] into n equal subintervals of length $\delta = (B - A)/n$. Let $k = k(n)$ be the largest integer such that Y meets the kth subinterval. Then there is a point $(g(n),h(n))$ such that

$$(g(n),h(n)) \in D,\ A + (k - 1)\delta \le f(g(n),h(n)),$$

but for all $(x,y) \in D$,

$$f(x,y) \le A + k\delta .$$

Now let n_1 be a positive infinite hyperinteger and let $\delta_1 = (B - A)/n_1$, $k_1 = k(n_1)$. Using the Solution Axiom we see that

$$(g(n_1),h(n_1)) \in D^*,\ A + (k_1 - 1)\delta_1 \le f(g(n_1),h(n_1)).$$

By another use of the Solution Axiom, for every point $(x_1,y_1) \in D^*$ we have

$$f(x_1,y_1) \le A + k_1\delta_1 .$$

Since D is closed and bounded, $(a,b) \in D$ where

$$a = st(g(n_1)), \ b = st(h(n_1)) .$$

By continuity of f,

$$f(a,b) = st\ f(g(n_1),h(n_1)) .$$

It follows that for any point $(x,y) \in D$,

$$f(x,y) \le st(A + k_1\delta_1) = st(A + (k_1 - 1)\delta_1) \le f(a,b),$$

so f has a maximum at (a,b). ⊣

The above proof differs from the proof of the one variable Extreme Value Theorem in Section 3C because this time we partitioned the range of f instead of the domain of f. The one variable proof can be generalized to the present case by using a rectangular grid to partition the domain of f.

A real point (a,b) is said to be a <u>critical point</u> of f if both partial derivatives of f are equal to zero,

$$f_x(a,b) = 0, \ f_y(a,b) = 0 .$$

CRITICAL POINT THEOREM

Suppose f is continuous on its domain D and f is differentiable at every interior point of D. If f has a maximum or minimum at a real point (a,b), then (a,b) is either a boundary point of D or a critical point of f.

<u>Proof</u> Suppose f has a maximum at an interior point (a,b). Then the one variable function $g(x) = f(x,b)$ has a maximum at the interior point a. By the one variable Critical Point Theorem, $g'(a) = f_x(a,b) = 0$. Similarly, $f_y(a,b) = 0$. ⊣

As in the one variable case, the Extreme Value Theorem and Critical Point Theorem lead to a method for finding the maxima and minima of a function with a bounded closed domain. We shall see that the method can often be applied for other domains as well. In <u>Elementary Calculus</u> we concentrated on <u>basic closed regions</u>, that is, sets D of the form

$$D = \{(x,y): a \le x \le b,\ g(x) \le y \le h(x)\}$$

where g and h are continuous and $g(x) \le h(x)$ for x in [a,b]. The following lemma ties the present approach to the more elementary treatment in <u>Elementary Calculus</u>.

LEMMA 3 Let D be a basic closed region in the plane.

(i) D is a bounded closed set.

(ii) The boundary of D is the set of all points of D on one of the curves

$$x = a,\ x = b,\ y = g(x),\ y = h(x)\ .$$

<u>Proof</u> (i) Since g and h are continuous, they have maxima and minima on [a,b], so D is bounded. Let $(x_1,y_1) \in D^*$ and let $x = st(x_1)$, $y = st(y_1)$. Then

$$a \le x_1 \le b,\ g(x_1) \le y_1 \le h(x_1)\ .$$

By continuity,

$$g(x) = st\ g(x_1),\ h(x) = st\ h(x_1)\ .$$

Therefore

$$a \le x \le b,\ g(x) \le y \le h(x),$$

so $(x,y) \in D$.

(ii) Obviously any point on one of the four curves is on the boundary of D. For example, if $g(x) = y$ then for any real $\varepsilon > 0$, $y - \varepsilon < g(x)$ and $(x,y-\varepsilon) \notin D$, so $N_\varepsilon(x,y)$ is not contained in D. Suppose $(x,y) \in D$ but (x,y) is not on one of the boundary curves. Then

$$a < x < b,\ g(x) < y < h(x)\ .$$

For any $(x_1,y_1) \approx (x,y)$ we have $a < x_1 < b$. Also, $g(x_1) < y_1 < h(x_1)$ because if, for example, $y_1 \le g(x_1)$, then

$$y = st(y_1) \le st(g(x_1)) = g(st(x_1)) = g(x)\ .$$

Thus $(x_1, y_1) \in D^*$ and D^* contains the monad of (x,y). It follows by Theorem 2 in Section 1D that (x,y) belongs to the interior of D. ⊣

In many applications, a function f is continuous on a basic closed region D and is differentiable and has only finitely many critical points on the interior of D. The maximum of f can usually be found as follows. First, evaluate f at each of its critical points. Second, on each of the four boundary curves of D find the maxima of f by eliminating one variable and using the one variable method. Finally, the largest of the values of f at the critical points and on the boundary curves must be the maximum value of f.

We now present a method for finding maxima and minima on open regions. A basic open region is a set $D \subseteq R^2$ of one of the forms

$$\{(x,y): x \in I, g(x) < y < h(x)\},$$
$$\{(x,y): x \in I, -\infty < y < h(x)\},$$
$$\{(x,y): x \in I, g(x) < y < \infty\},$$
$$\{(x,y): x \in I, -\infty < y < \infty\},$$

where I is an open interval, f and g are continuous real functions on I, and $g(x) < h(x)$ for all $x \in I$.

A basic closed hyperreal region is a set $E \subseteq (R^*)^2$ of hyperreal points of the form

$$E = \{(x,y): a \le x \le b, f(x) + A \le y \le g(x) + B\}$$

where a, b, A, B are hyperreal numbers and f, g are real functions continuous on an interval I such that $a, b \in I^*$.

A hyperreal cover of a basic open region D is a basic closed hyperreal region E obtained as follows. Let ε be a positive infinitesimal and H a positive infinite hyperreal number. If

$$D = \{(x,y) \in R^2: a < x < b, g(x) < y < h(x)\},$$

then

$$E = \{(x,y) \in R^{*2}: a + \varepsilon \le x \le b - \varepsilon, \quad g(x) + \varepsilon \le y \le h(x) - \varepsilon\}.$$

If

$$D = \{(x,y) \in R^2: a < x < \infty, g(x) < y < \infty\},$$

then

$$E = \{(x,y) \in R^{*2}: a + \varepsilon \le x \le H,\ g(x) + \varepsilon \le y \le H\} .$$

The other cases are similar.

LEMMA 4 If E is a hyperreal cover of a basic open region D, then $D \subsetneq E \subsetneq D^*$.

This lemma is immediate from the definition.

The standard calculus course often gives the second derivative test for local maxima, minima, and saddle points, but no practical test for global maxima and minima. The following method for finding global maxima and minima for open regions is presented in Elementary Calculus. It is an application of hyperreal numbers to an elementary problem about real functions. We know from the Critical Point Theorem that if a maximum exists it must occur at a critical point, but it is often hard to determine whether or not a maximum exists. The following result is a practical criterion for the existence of a maximum.

By a maximum critical point of $f(x,y)$ we mean a critical point (x_0,y_0) such that for every other critical point (x_1,y_1), $f(x_0,y_0) \ge f(x_1,y_1)$. Obviously, if $f(x,y)$ has at least one but only finitely many critical points, it has a maximum critical point.

THEOREM 5

Suppose a real function $f(x,y)$ is defined on a basic open region D, has partial derivatives at every point of D, and has a maximum critical point (x_0,y_0). Let E be a hyperreal cover of D. Then

(i) If $f(x_0,y_0) \ge f(x,y)$ for all (x,y) on the boundary of E, then f has a maximum in D at (x_0,y_0).

(ii) If $f(x_0,y_0) < f(x,y)$ for some (x,y) on the boundary of E, then f has no maximum in D.

A similar result holds for minima.

Proof We give the proof in the case that D has the form

$$D = \{(x,y): a < x < \infty,\ g(x) < y < \infty\} .$$

Then

$$E = \{(x,y): a + \varepsilon \le x \le H,\ g(x) + \varepsilon \le y \le H\}$$

where ε is positive infinitesimal and H is positive infinite. We first prove (ii). Suppose f has a maximum in D. By the Critical Point Theorem, the maximum must occur at (x_0,y_0). Then by Theorem 1, $f(x_0,y_0) \ge f(x,y)$ for all $(x,y) \in D^*$. Since $E \subseteq D^*$, $f(x_0,y_0) \ge f(x,y)$ for all (x,y) on the boundary of E. The proof of (i) uses the Partial Solution Theorem. For each positive real ε_0 and H_0 let $E(\varepsilon_0,H_0)$ be the basic closed region

$$E(\varepsilon_0,H_0) = \{(x,y): a + \varepsilon_0 \le x \le H_0,\ g(x) + \varepsilon_0 \le y \le H_0\}$$

and let $\partial E(\varepsilon_0,H_0)$ be the boundary of $E(\varepsilon_0,H_0)$. Then

$$E(\varepsilon_0,H_0) \subseteq D \subseteq E \subseteq D^* .$$

The relations $(x,y) \in E(\varepsilon_0,H_0)$ and $(x,y) \in \partial E(\varepsilon_0,H_0)$ can both be expressed by systems of formulas. By the Extreme Value and Critical Point Theorems, if $(x_0,y_0) \in E(\varepsilon_0,H_0)$ then in $E(\varepsilon_0,H_0)$, f has a maximum either at (x_0,y_0) or on the boundary of $E(\varepsilon_0,H_0)$. Thus every real solution of

$$\begin{aligned} (x_0,y_0) \in E(\varepsilon_0,H_0),\ (s,t) \in E(\varepsilon_0,H_0), \\ f(x_0,y_0) < f(s,t) \end{aligned} \tag{1}$$

is a partial solution of

$$(u,v) \in \partial E(\varepsilon_0,H_0),\ f(x_0,y_0) < f(u,v) . \tag{2}$$

Suppose f does not have a maximum in D at (x_0,y_0). Then $f(x_0,y_0) < f(s,t)$ for some real point $(s,t) \in D$. Since $D \subseteq E = E(\varepsilon,H)$, the system of formulas (1) holds for (ε,H). By the Partial Solution Theorem there is a point (u,v) such that (2) holds. That is, (u,v) is on the boundary of E and $f(x_0,y_0) < f(u,v)$. This proves (i). ⊣

11D SECOND PARTIAL DERIVATIVES (§11.8)

We shall give a hyperreal proof of the equality of the mixed second partial derivatives. The hypothesis we need is the uniform differentiability of both first partial derivatives. (In Elementary Calculus we used the stronger hypothesis that both first partial derivatives are smooth.) The proof uses the following form of the Hyperreal Mean Value Theorem for two variables.

HYPERREAL MEAN VALUE THEOREM

Suppose $f(x,y)$ is a real function whose partial derivative $f_x(x,y)$ exists on an open rectangle D. Then for every pair of hyperreal points (x,y) and $(x + \Delta x,y)$ in D^* with $\Delta x \neq 0$, there is a point t between x and $x + \Delta x$ such that

$$\frac{f(x + \Delta x,y) - f(x,y)}{\Delta x} = f_x(t,y) .$$

A similar result holds for the other partial derivative f_y.

Proof By the Real Mean Value Theorem in one variable, every real solution of

$$(x,y) \in D, \quad (x + \Delta x,y) \in D, \quad \Delta x > 0 \tag{1}$$

is a partial real solution of

$$\frac{f(x + \Delta x,y) - f(x,y)}{\Delta x} = f_x(t,y), \quad x < t < x + \Delta x . \tag{2}$$

The result now follows from the Partial Solution Theorem. ┤

THEOREM 1

Suppose both partial derivatives $\partial f/\partial x$ and $\partial f/\partial y$ of a real function $f(x,y)$ are uniformly differentiable at a real point (a,b). Then the mixed second partial derivatives are equal at (a,b),

$$\frac{\partial^2 f}{\partial x \partial y}(a,b) = \frac{\partial^2 f}{\partial y \partial x}(a,b) .$$

Proof Let Δx and Δy be nonzero infinitesimals. Since f_x and f_y are uniformly differentiable at (a,b), they exist on a real neighborhood of (a,b). Let

$$D = f(a + \Delta x, b + \Delta y) - f(a + \Delta x, b) - f(a, b + \Delta x) - f(a,b).$$

By the Hyperreal Mean Value Theorem applied to $g(x,\Delta y) = f(x, b + \Delta y) - f(x,b)$, we have

$$\frac{D}{\Delta x} = \frac{[f(a + \Delta x, b + \Delta y) - f(a + \Delta x, b)] - [f(a, b + \Delta y) - f(a,b)]}{\Delta x}$$

$$= \frac{\partial f}{\partial x}(t, b + \Delta y) - \frac{\partial f}{\partial x}(t,b)$$

for some t between a and $a + \Delta x$. Also,

$$\frac{D}{\Delta y} = \frac{[f(a + \Delta x, b + \Delta y) - f(a, b + \Delta y)] - [f(a + \Delta x, b) - f(a,b)]}{\Delta y}$$

$$= \frac{\partial f}{\partial y}(a + \Delta x, u) - \frac{\partial f}{\partial y}(a,u)$$

for some u between b and $b + \Delta y$. By uniform differentiability,

$$\frac{D}{\Delta x \Delta y} = \frac{\frac{\partial f}{\partial x}(t, b + \Delta y) - \frac{\partial f}{\partial x}(t,b)}{\Delta y} \approx \frac{\partial^2 f}{\partial y \partial x}(a,b)$$

and

$$\frac{D}{\Delta x \Delta y} = \frac{\frac{\partial f}{\partial y}(a + \Delta x, u) - \frac{\partial f}{\partial y}(a,u)}{\Delta x} \approx \frac{\partial^2 f}{\partial x \partial y}(a,b) .$$

Therefore

$$\frac{\partial^2 f}{\partial y \partial x}(a,b) = \frac{\partial^2 f}{\partial x \partial y}(a,b) . \qquad \dashv$$

11E LINE INTEGRALS (§11.10)

DEFINITION

A <u>smooth curve</u> (or directed curve) <u>from</u> A <u>to</u> B is a real parametric curve

$$C: 0 \le s \le L, \; x = g(s), \; y = h(s)$$

where $g'(s)$ and $h'(s)$ are continuous,

$$A = (g(0),h(0)), \; B = (g(L),h(L)),$$

and s is the length of the curve from A to (x,y). A and B are called the <u>initial point</u> and <u>terminal point</u> respectively.

In <u>Elementary Calculus</u> the line integral was defined as an ordinary definite integral as follows.

DEFINITION

Let $\vec{F}(x,y) = P(x,y)\vec{i} + Q(x,y)\vec{j}$ be a continuous real vector valued function on an open rectangle D containing a smooth curve C. The <u>line integral</u> of $\vec{F}$ along C, denoted by

$$\int_C \vec{F}\cdot d\vec{S} = \int_C P\cdot dx + Q\cdot dy \; ,$$

is defined as the definite integral

$$\int_0^L (P \frac{dx}{ds} + Q \frac{dy}{ds})ds$$

where $x = g(s)$ and $y = h(s)$.

In this section we shall give a hyperreal characterization of the line integral.

Let n be a positive integer and let $\Delta s = L/n$. For $k = 0,\ldots,n - 1$ let

$$s_k = k\Delta s, \; x_k = g(s_k), \; y_k = h(s_k),$$

$$\Delta x_k = x_{k+1} - x_k, \; \Delta y_k = y_{k+1} - y_k,$$

$$\Delta\vec{S}_k = \Delta x_k\vec{i} + \Delta y_k\vec{j} \; .$$

The finite Riemann sum along C,

$$\sum_C \vec{F}\cdot\Delta\vec{S} = \sum_C P\Delta x + Q\Delta y,$$

is defined as the sum

$$\sum_{k=0}^{n} \vec{F}(x_k,y_k)\cdot\Delta\vec{S}_k = \sum_C P(x_k,y_k)\Delta x_k + Q(x_k,y_k)\Delta y_k \ .$$

This sum corresponds to a polygonal line connecting n + 1 points along C. It depends only on n, so we may write

$$I(n) = \sum_C \vec{F}\cdot\Delta\vec{S} \ .$$

Now let H be a positive infinite hyperinteger. The value I(H) is called an <u>infinite Riemann sum</u> along C and is denoted by

$$I(H) = \sum_C \vec{F}\cdot d\vec{S} = \sum_C P\cdot dx + Q\cdot dy \ .$$

THEOREM 1

Let $\vec{F}(x,y)$ be a continuous real vector valued function on an open rectangle D containing a smooth curve C. Let H be a positive infinite hyperinteger. Then the line integral along C is equal to the standard part of the infinite Riemann sum along C,

$$\int_C \vec{F}\cdot d\vec{S} = st(\sum_C \vec{F}\cdot d\vec{S}) \ ,$$

or

$$\int_C P\cdot dx + Q\cdot dy = st(\sum_C P\cdot dx + Q\cdot dy) \ .$$

<u>Proof</u> First consider a positive integer n and form the finite Riemann sum

$$\sum_C P\Delta x + Q\Delta y = \sum_{k=0}^{n} P(x_k,y_k)\Delta x_k + Q(x_k,y_k)\Delta y_k \ .$$

Let r be a positive real number. If the Riemann sum differs from the line integral by at least r, then on one of the subintervals the Riemann sum differs from the line integral by at least r/n. That is, every real solution of

$$n \in Z,\ 0 < n,\ \left|\sum_C P\Delta x + Q\Delta y - \int_C P\cdot dx + Q\cdot dy\right| \geq r \qquad (1)$$

is a partial real solution of

$$\Delta s = L/n,\ 0 \le s \le s + \Delta s \le L,\ x = g(s),\ y = h(s),$$
$$\Delta x = g(s + \Delta s) - g(s),\ \Delta y = h(s + \Delta s) - h(s), \tag{2}$$
$$\left| P(x,y)\Delta x + Q(x,y)\Delta y - \int_s^{s+\Delta s} (Pg' + Qh')ds \right| \ge r/n\ .$$

By the Partial Solution Theorem every hyperreal solution of (1) is a partial solution of (2). Now form the infinite Riemann sum $\sum_C P\cdot dx + Q\cdot dy$ with respect to H. Assume that

$$\left| \sum_C P\cdot dx + Q\cdot dy - \int_C P\cdot dx + Q\cdot dy \right| \ge r\ . \tag{3}$$

Let $\Delta_1 s = L/H$. Then there exist $0 \le s_1 \le s_1 + \Delta_1 s \le L$ such that, putting

$$x_1 = g(s_1),\ y_1 = h(s_1),\ \Delta_1 x = g(s_1 + \Delta_1 s) - g(s_1),$$
$$\Delta_1 y = h(s_1 + \Delta_1 s) - h(s_1),$$

we have

$$\left| P(x_1,y_1)\Delta_1 x + Q(x_1,y_1)\Delta_1 y - \int_{s_1}^{s_1+\Delta_1 s} (Pg' + Qh')ds \right| \ge \frac{r}{H}\ .$$

Dividing by $\Delta_1 s$,

$$\left| P(x_1,y_1)\frac{\Delta_1 x}{\Delta_1 s} + Q(x_1,y_1)\frac{\Delta_1 y}{\Delta_1 s} - \frac{\int_{s_1}^{s_1+\Delta_1 s} (Pg' + Qh')ds}{\Delta_1 s} \right| \ge \frac{r}{L}\ . \tag{4}$$

Since $x = g(s)$, $y = h(s)$, and $\int_0^s (Pg' + Qh')ds$ are continuously differentiable functions of s, we have

$$\frac{\Delta_1 x}{\Delta_1 s} \approx g'(s_1),\ \frac{\Delta_1 y}{\Delta_1 s} \approx h'(s_1),$$

$$\frac{\int_{s_1}^{s_1+\Delta s_1} (Pg' + Qh')ds}{\Delta_1 s} \approx P(x_1,y_1)g'(s_1) + Q(x_1,y_1)h'(s_1)\ .$$

But then

$$P(x_1,y_1)\frac{\Delta_1 x}{\Delta_1 s} + Q(x_1,y_1)\frac{\Delta_1 y}{\Delta_1 s} \approx \frac{\int_{s_1}^{s_1+\Delta_1 s}(Pg' + Qh')ds}{\Delta_1 s} ,$$

contradicting (4). We conclude that (3) fails for every positive real number r. It follows that

$$\sum_C P\cdot dx + Q\cdot dy \approx \int_C P\cdot dx + Q\cdot dy . \qquad \dashv$$

MULTIPLE INTEGRATION 12

Double and triple integrals are developed in Chapter 12 of Elementary Calculus. To simplify our notation we shall concentrate entirely on double integrals here. However, everything we do in this chapter can be readily generalized to triple integrals.

12A DOUBLE INTEGRALS (§12.1, §12.2)

We shall define the double integral of a continuous real function $f(x,y)$ and state the basic results. Most of the proofs are similar to corresponding proofs for the single integral in Chapter 4 and will be omitted here.

We assume throughout this section that $f(x,y)$ is a real function continuous on an open set D_0. We shall consider basic closed subregions D of D_0. The analogue of an area function for a real function of one variable is a volume function.

A volume function for f is a function $B(D)$ from the set of basic closed regions $D \subseteq D_0$ into the real numbers which has the following properties:

ADDITION PROPERTY If $D = D_1 \cup D_2$ where

$$D_1 = \{(x,y) \in D: x \leq c\},$$

$$D_2 = \{(x,y) \in D: x \geq c\},$$

then $B(D) = B(D_1) + B(D_2)$.

If $D = D_1 \cup D_2$ where

$$D_1 = \{(x,y) \in D: y \le k(x)\} ,$$

$$D_2 = \{(x,y) \in D: y \ge k(x)\} ,$$

then $B(D) = B(D_1) + B(D_2)$.

CYLINDER PROPERTY If $f(x,y)$ has minimum value m and maximum value M on D, and A is the area of D, then

$$mA \le B(D) \le MA .$$

The Addition Property states that if we divide D into two basic closed regions D_1 and D_2 with either a vertical line $x = c$ or a continuous curve $y = k(x)$, then $B(D) = B(D_1) + B(D_2)$. The intuitive notion of the volume of the solid above the region D between the plane $z = 0$ and the surface $z = f(x,y)$ has the Addition and Cylinder Properties. It will turn out that the double integral is the unique volume function for f. Our first step in constructing the double integral is the finite double Riemann sum.

Let

$$D = \{(x,y): a_1 \le x \le a_2, b_1(x) \le y \le b_2(x)\}$$

be a basic closed region included in D_0, and let

$$B_1 = \text{minimum value of } b_1(x),\ a_1 \le x \le a_2,$$

$$B_2 = \text{maximum value of } b_2(x),\ a_1 \le x \le a_2.$$

The rectangle

$$[a_1,a_2] \times [B_1,B_2]$$

is called the <u>circumscribed rectangle</u> of D. Let Δx and Δy be positive real numbers and divide the intervals $[a_1,a_2]$ and $[B_1,B_2]$ into subintervals of length Δx and Δy respectively,

$$x_0 = a_1,\ x_1 = a_1 + \Delta x, \ldots, x_m = a_1 + m\Delta x$$

where $x_m < a_2 \le x_m + \Delta x$,

$$y_0 = B_1,\ y_1 = B_1 + \Delta y, \ldots, y_n = B_1 + n\Delta y$$

where $y_n < B_2 \le y_n + \Delta y$. If Δx does not evenly divide $a_2 - a_1$, the last subinterval $[x_m, a_2]$ will have length less than Δx and will be covered by the interval $[x_m, x_m + \Delta x]$. Thus the circumscribed rectangle $[a_1, a_2] \times [B_1, B_2]$ of D is covered by a grid of Δx by Δy rectangles.

DEFINITION

The <u>finite double Riemann sum</u> of f over D is defined as

$$\sum\sum_D f(x,y)\Delta x\Delta y = \sum_{k=0}^{m} \sum_{\ell=0}^{n} \{f(x_k, y_\ell)\Delta x\Delta y\colon (x_k, y_\ell) \in D\}.$$

Thus $\sum\sum_D f(x,y)\Delta x\Delta y$ is the sum of the volumes of the rectangles of base $\Delta x\Delta y$ and height $f(x_k, y_\ell)$ such that the point (x_k, y_ℓ) belongs to D. For a given D, the finite double Riemann sum $\sum\sum_D f(x,y)\Delta x\Delta y$ is a real function of two variables Δx and Δy, and is defined for all $\Delta x > 0$, $\Delta y > 0$. By the Function Axiom it is also defined for all hyperreal $\Delta x > 0$, $\Delta y > 0$.

DEFINITION

If dx and dy are positive infinitesimals, the hyperreal number

$$\sum\sum_D f(x,y)\,dxdy$$

is called the <u>infinite Riemann sum</u> of f over D with respect to dx and dy.

LEMMA 1 For any positive infinitesimals dx and dy, the infinite Riemann sum $\sum\sum_D f(x,y)dxdy$ is a finite hyperreal number.

DEFINITION

Let dx and dy be positive infinitesimals. The <u>double integral</u> of f over D with respect to dx and dy is the standard part of the infinite Riemann sum,

$$\int\int_D f(x,y)\,dxdy = \sum\sum_D f(x,y)\,dxdy\ .$$

THEOREM 2

The value of the double integral of f over D does not depend on dx and dy. That is, if dx, d_1x, dy, d_1y are positive infinitesimals, then

$$\iint_D f(x,y)dxdy = \iint_D f(x,y)d_1xd_1y \ .$$

Hereafter we shall write dA = dxdy and denote the double integral by

$$\iint_D f(x,y)dxdy = \iint_D f(x,y)dA \ .$$

THEOREM 3

The double integral is the unique volume function for f.

This theorem justifies the definition of the volume V of the solid over the region D above the plane z = 0 and below the surface z = f(x,y) as

$$V = \iint_D f(x,y)dA \ .$$

It follows from the Cylinder Property that the area of a region D is equal to the definite integral

$$A = \iint_D 1 \ dA \ .$$

The following theorem provides the basic tool for computing double integrals.

ITERATED INTEGRAL THEOREM

Let D be the basic closed region

$$D = \{(x,y): a_1 \le x \le a_2, \ b_1(x) \le y \le b_2(x)\} \ .$$

Then the double integral of f over D is equal to the iterated integral

$$\iint_D f(x,y)dA = \int_{a_1}^{a_2} \int_{b_1(x)}^{b_2(x)} f(x,y)dydx \ .$$

Proof Let

$$B(D) = \int_{a_1}^{a_2} \int_{b_1(x)}^{b_2(x)} f(x,y)\,dydx \,.$$

Using properties of the single integral, a simple computation shows that B(D) is a volume function for f. ⊣

COROLLARY 3 Suppose D is a basic closed region in both the (x,y) plane and the (y,x) plane, that is,

$$D = \{(x,y): a_1 \le x \le a_2,\ b_1(x) \le y \le b_2(x)\} \,,$$

$$D = \{(x,y): B_1 \le y \le B_2,\ c_1(y) \le x \le c_2(y)\} \,.$$

Then the two iterated integrals are equal,

$$\int_{a_1}^{a_2} \int_{b_1(x)}^{b_2(x)} f(x,y)\,dydx = \int_{B_1}^{B_2} \int_{c_1(y)}^{c_2(y)} f(x,y)\,dxdy \,.$$

Proof The definition of the double Riemann sum is symmetric between x and y, so the double integral in the sense of the (x,y) plane equals the double integral in the sense of the (y,x) plane. By the Iterated Integral Theorem,

$$\int_{a_1}^{a_2} \int_{b_1(x)}^{b_2(x)} f(x,y)\,dydx = \iint_D f(x,y)\,dA$$

$$= \int_{B_1}^{B_2} \int_{c_1(y)}^{c_2(y)} f(x,y)\,dxdy \,.$$ ⊣

12B GREEN'S THEOREM (§11.11, §12.8)

We shall prove Green's Theorem for basic closed regions with smooth boundaries, using the Infinite Sum Theorem for one variable and a lemma on partial derivatives of integrals. This lemma was stated in Section 11.11 of Elementary Calculus, where it was used to prove the Path Independence Theorem for line integrals. We sketched a proof of the lemma in Elementary

Calculus and will give a detailed proof here. In Section 12.8 of Elementary Calculus, Green's Theorem was stated for basic closed regions but proved only for rectangles.

LEMMA Suppose $P(x,y)$ is a smooth function on an open rectangle D containing the point (a,b). Then

$$\frac{\partial}{\partial x}\int_a^x P(t,y)\,dt = P(x,y),$$

$$\frac{\partial}{\partial y}\int_a^x P(t,y)\,dt = \int_a^x \frac{\partial P}{\partial y}(t,y)\,dt .$$

Proof The first formula follows at once from the Second Fundamental Theorem of Calculus. We prove the second formula. Let

$$F(x,y) = \int_a^x P(t,y)\,dt, \quad (x,y) \in D.$$

Any real solution of

$$(x,y) \in D, \quad (x,y+\Delta y) \in D, \quad \Delta y \neq 0 \tag{1}$$

is a solution of

$$\frac{F(x,y+\Delta y) - F(x,y)}{\Delta y} = \int_a^x \frac{P(t,y+\Delta y) - P(t,y)}{\Delta y}\,dt . \tag{2}$$

Moreover, by the Mean Value Theorem for integrals, any real solution of (1) is a partial real solution of

$$\int_a^x \left[\frac{P(t,y+\Delta y) - P(t,y)}{\Delta y} - \frac{\partial P}{\partial y}(t,y)\right]dt$$
$$= (x-a)\left[\frac{P(u,y+\Delta y) - P(u,y)}{\Delta y} - \frac{\partial P}{\partial y}(u,y)\right], \tag{3}$$

u between a and x.

Now let $(x,y) \in D$ and $\Delta_1 y \underset{\neq}{\approx} 0$. Then $(x, y+\Delta_1 y) \in D^*$. By the Partial Solution Theorem, (2) and (3) hold for some hyperreal u_1. Since P is smooth, it is uniformly differentiable on D, whence

$$\frac{P(u_1, y+\Delta_1 y) - P(u_1,y)}{\Delta_1 y} \approx \frac{\partial P}{\partial y}(u_1,y) .$$

Then by (3),

$$\int_a^x \frac{P(t,y+\Delta_1 y) - P(t,y)}{\Delta_1 y}\,dt \approx \int_a^x \frac{\partial P}{\partial t}(t,y)dt\ .$$

Finally, by (2),

$$\frac{F(x,y+\Delta_1 y) - F(x,y)}{\Delta_1 y} = \int_a^x \frac{P(t,y+\Delta_1 y) - P(x,y)}{\Delta_1 y}\,dt$$

$$\approx \int_a^x \frac{\partial P}{\partial y}(t,y)dt\ ,$$

so

$$\frac{\partial F}{\partial y}(x,y) = \int_a^x \frac{\partial P}{\partial y}(t,y)dt\ . \qquad \dashv$$

Given a basic closed region

$$D = \{(x,y): a_1 \le x \le a_2,\ b_1(x) \le y \le b_2(x)\}\ ,$$

we denote the boundary of D by ∂D. The line integral of $P\cdot dx + Q\cdot dy$ around ∂D in the counterclockwise direction is denoted by

$$\oint_{\partial D} P\cdot dx + Q\cdot dy\ .$$

GREEN'S THEOREM

Suppose $P(x,y)$ and $Q(x,y)$ are smooth functions on an open set containing a basic closed region D with smooth upper and lower boundaries. Then

$$\oint_{\partial D} P\cdot dx + Q\cdot dy = \iint_D \left(\frac{\partial Q}{\partial x} - \frac{\partial P}{\partial y}\right)dA\ .$$

<u>Proof</u> For $a_1 \le u < v \le a_2$, let

$$B(u,v) = \oint_{\partial D_1} P\cdot dx + Q\cdot dy$$

where D_1 is the part of D with $u \le x \le v$. Then B has the Addition Property, because if D_1 and D_2 meet only on a common vertical boundary, the parts of the line integrals on the common boundary cancel. Our plan is to show that for any infinitesimal subinterval $[x, x + \Delta x]^*$ of $[a_1, a_2]^*$,

$$\frac{\Delta B}{\Delta x} \approx P(x, b_1(x)) - P(x, b_2(x)) + \int_{b_1(x)}^{b_2(x)} \frac{\partial Q}{\partial x}\, dy \,. \qquad (4)$$

After (4) is verified, the proof is completed as follows. By the Infinite Sum Theorem,

$$B(a_1, a_2) = \int_{a_1}^{a_2} \left[P(x, b_1(x)) - P(x, b_2(x)) + \int_{b_1(x)}^{b_2(x)} \frac{\partial Q}{\partial x}\, dy \right] dx.$$

By the Fundamental Theorem of Calculus,

$$\int_{b_1(x)}^{b_2(x)} -\frac{\partial P}{\partial y}\, dy = P(x, b_1(x)) - P(x, b_2(x))$$

for each real $x \in [a_1, a_2]$. Therefore

$$\oint_{\partial D} P \cdot dx + Q \cdot dy = B(a_1, a_2)$$

$$= \int_{a_1}^{a_2} \int_{b_1(x)}^{b_2(x)} \left(\frac{\partial Q}{\partial x} - \frac{\partial P}{\partial y} \right) dy dx = \iint_D \left(\frac{\partial Q}{\partial x} - \frac{\partial P}{\partial y} \right) dA \,.$$

It remains to prove (4). First let $[x, x + \Delta x]$ be a real subinterval of $[a_1, a_2]$, that is,

$$a_1 \le x < x + \Delta x \le a_2 \,. \qquad (5)$$

Then

$$\frac{\Delta B}{\Delta x} = \frac{1}{\Delta x} \int_x^{x+\Delta x} [P(t, b_1(t)) + Q(t, b_1(t)) b_1'(t)] dt$$

$$+ \frac{1}{\Delta x} \int_{b_1(x+\Delta x)}^{b_2(x+\Delta x)} Q(x + \Delta x, y) dy$$

$$- \frac{1}{\Delta x} \int_x^{x+\Delta x} [P(t,b_2(t)) + Q(t,b_2(t))b_2'(t)]dt$$

$$- \frac{1}{\Delta x} \int_{b_1(x)}^{b_2(x)} Q(x,y)dy .$$

We rewrite this equation in the form

$$\frac{\Delta B}{\Delta x} = \frac{1}{\Delta x} \int_x^{x+\Delta x} [P(t,b_1(t)) - P(t,b_2(t))]dt \tag{6}$$

$$+ \frac{1}{\Delta x} \int_x^{x+\Delta x} [Q(t,b_1(t))b_1'(t) - Q(t,b_2(t))b_2'(t)]dt$$

$$+ \int_{b_1(x)}^{b_2(x)} \frac{Q(x + \Delta x,y) - Q(x,y)}{\Delta x} dy$$

$$+ \frac{1}{\Delta x} \int_{b_1(x+\Delta x)}^{b_1(x)} Q(x+\Delta x,y)dy + \frac{1}{\Delta x} \int_{b_2(x)}^{b_2(x+\Delta x)} Q(x + \Delta x,y)dy.$$

Now let $[x,x + \Delta x]^*$ be an infinitesimal subinterval of $[a_1,a_2]$. Then $(x,x + \Delta x)$ is a hyperreal solution of (5). By the Solution Axiom it is also a solution of (6). By the Lemma,

$$\frac{\partial}{\partial t} \int_{b_1(x)}^{b_2(x)} Q(t,y)dy = \int_{b_1(x)}^{b_2(x)} \frac{\partial Q}{\partial t} (t,y)dy . \tag{7}$$

The right side of (7) is continuous in t and x, so $\int_{b_1(x)}^{b_2(x)} Q(t,y)dy$ is uniformly differentiable. Therefore when $x = t$ we have

$$\frac{1}{\Delta x} \int_{b_1(x)}^{b_2(x)} [Q(x + \Delta x,y) - Q(x,y)]dy$$

$$\approx \int_{b_1(x)}^{b_2(x)} \frac{\partial Q}{\partial x} (x,y)dy .$$

Applying the Hyperreal Mean Value Theorem to (6), we obtain

$$\frac{\Delta B}{\Delta x} \approx P(x,b_1(x)) - P(x,b_2(x)) + Q(x,b_1(x))b_1'(x) - Q(x,b_2(x))b_2'(x)$$

$$+ \int_{b_1(x)}^{b_2(x)} \frac{\partial Q}{\partial x}(x,y)dy - Q(x + \Delta x,b_1(x))\frac{\Delta b_1(x)}{\Delta x} - Q(x + \Delta x,b_2(x))\frac{\Delta b_2(x)}{\Delta x}$$

$$\approx P(x,b_1(x)) - P(x,b_2(x)) + \int_{b_1(x)}^{b_2(x)} \frac{\partial Q}{\partial x}(x,y)\,dy \ .$$

This completes the proof of (4). ⊣

12C INFINITE SUM THEOREM FOR TWO VARIABLES (§12.3)

As in the case of single integrals, the Infinite Sum Theorem can be used as the basis for applications of the double integral.

DEFINITION

An <u>element of area</u> is a hyperreal rectangle ΔD whose sides are infinitesimal and parallel to the x and y axes. Given an element of area ΔD we let

$$(x,y) = \text{lower left corner of } \Delta D,$$
$$\Delta x,\ \Delta y = \text{dimensions of } \Delta D,$$
$$\Delta A = \Delta x \Delta y = \text{area of } \Delta D.$$

INFINITE SUM THEOREM

Let $h(x,y)$ be a real function continuous on an open set D_0 and let $B(D)$ be a function from basic closed regions $D \subseteq D_0$ to real numbers. Let Δx and Δy be positive infinitesimals. Assume that:

(i) B has the Addition Property;

(ii) $B(D) \geq 0$ for all D;

(iii) For every element of area $\Delta D \subseteq D_0^*$ with dimensions Δx and Δy,

$$B(\Delta D) \approx h(x,y)\Delta x \Delta y \quad (\text{compared to } \Delta x \Delta y).$$

Then for every basic closed region $D \subseteq D_0$,

$$B(D) = \iint_D h(x,y)\,dA \ .$$

A standard form of the Infinite Sum Theorem is proved in the book of Buck in the section on set functions. In <u>Elementary Calculus</u> we stated a weaker form of the Infinite Sum Theorem, where (iii) is assumed for all elements of area rather than only for elements of area with fixed dimensions Δx and Δy. The proof there was given only in the case that D is a rectangle. We give the proof in general here. The general proof is much longer than the proof for D a rectangle, and uses the following four classical lemmas.

LEMMA 1 Let D be a closed and bounded subset of D_0. There is a positive real number ε such that the ε-neighborhood of every point of D is included in D_0.

<u>Proof</u> Suppose that for each real $\varepsilon > 0$ there are points $P \in D$ and $Q \notin D_0$ such that $|P - Q| < \varepsilon$. Then by the Partial Solution Theorem there are points $P_1 \in D^*$, $Q_1 \notin D_0^*$ with $P_1 \approx Q_1$. Since D is closed and bounded, P_1 has a standard part P_0, and $P_0 \in D$. Thus $P_0 \in D_0$ and $Q_1 \approx P_0$, and since D_0 is open, $Q_1 \in D_0^*$. This contradiction completes the proof. ⊣

LEMMA 2 Suppose a real function f is continuous on D_0. Then for every basic closed region $D \subsetneq D_0$ there is a basic open region D_1 such that $D \subsetneq D_1 \subsetneq D_0$ and f is bounded on D_1.

<u>Proof</u> Let

$$D = \{(x,y): a_1 \le x \le a_2,\ b_1(x) \le y \le b_2(x)\}.$$

By Lemma 1 there is a real $\varepsilon > 0$ such that the ε-neighborhood of any point of D is included in D_0. Therefore the basic closed region

$$D_2 = \{(x,y): a_1 - \varepsilon \le x \le a_2 + \varepsilon,\ b_1(x) - \varepsilon \le y \le b_2(x) + \varepsilon\}$$

is included in D_0. By the Extreme Value Theorem, f has a maximum and minimum value in D_2, so f is bounded in D_2. Let D_1 be the interior of D_2,

$$D_1 = \{(x,y): a_1 - \varepsilon < x < a_2 + \varepsilon,\ b_1(x) - \varepsilon < y < b_2(x) + \varepsilon\}.$$

Then D_1 is open, $D \subsetneq D_1 \subsetneq D_0$, and f is bounded on D_1.

LEMMA 3 Suppose the function $B(D)$ on basic closed regions D has the Addition Property and $B(D) \geq 0$ for all D. Then whenever $D_1 \subseteq D_2$, $B(D_1) \leq B(D_2)$.

Proof By extending the left and right boundaries of the subregion D_1 to the upper and lower boundaries of D_2, we may partition D_2 into five basic closed regions

$$D_2 = D_1 \cup E_1 \cup E_2 \cup E_3 \cup E_4$$

which meet only on boundaries. By the Addition Property,

$$B(D_2) = B(D_1) + B(E_1) + B(E_2) + B(E_3) + B(E_4).$$

Since each term on the right is ≥ 0, $B(D_1) \leq B(D_2)$. ⊣

LEMMA 4 Let D be a basic closed region. For every real $\varepsilon > 0$ there is a real $\delta > 0$ such that for every partition of the circumscribed rectangle of D into subrectangles of length and width less than δ, the total area of the subrectangles which meet the boundary of D is less than ε.

Proof Let

$$D = \{(x,y): a_1 \leq x \leq a_2,\ b_1(x) \leq y \leq b_2(x)$$

and let $[a_1,a_2] \times [B_1,B_2]$ be the circumscribed rectangle of D. Let ε_1 be a positive real number. Since $b_1(x)$ and $b_2(x)$ are uniformly continuous on $[a_1,a_2]$, there is a real $\delta > 0$ such that whenever $x_1,x_2 \in [a_1,a_2]$ and $|x_2 - x_1| < \delta$, we have $\delta < \varepsilon_1$ and

$$|b_1(x_2) - b_1(x_1)| < \varepsilon_1, \ |b_2(x_2) - b_2(x_1)| < \varepsilon_1 .$$

Partition the circumscribed rectangle into a grid of Δx by Δy subrectangles where $0 < \Delta x < \delta$, $0 < \Delta y < \delta$. Each vertical boundary of D is covered by at most two columns of subrectangles and thus has total area less than $2\varepsilon_1(B_2 - B_1)$. Since $b_1(x)$ and $b_2(x)$ change by less than ε_1 over an interval of length Δx, each of the upper and lower boundaries of D is covered by a set of subrectangles of total area less than $2\varepsilon_1(a_2 - a_1)$. Therefore the set of subrectangles which meet the boundary of D has total area less than

$$4\varepsilon_1[(B_2 - B_1) + (a_2 - a_1)] .$$

Thus a value of δ corresponding to

$$\varepsilon_1 = \frac{\varepsilon}{4[(B_2 - B_1) + (a_2 - a_1)]}$$

has the required property. ⊣

Proof of the Infinite Sum Theorem Let D be the basic closed region

$$D = \{(x,y): a_1 \le x \le a_2, b_1(x) \le y \le b_2(x)\}$$

where $D \subseteq D_0$. We wish to show that $B(D) = \iint_D h(x,y)dA$. By Lemma 2, we may assume without loss of generality that h is bounded on D_0. $h(x,y) \ge 0$ on D_0 because of (i) and (iii). Thus for some real M, $0 \le h(x,y) \le M$ on D_0. By Lemma 1 there is a real $\varepsilon > 0$ such that the ε-neighborhood of every point in D is included in D_0. Let $[a_1,a_2] \times [B_1,B_2]$ be the circumscribed rectangle of D. Consider real numbers ΔX, ΔY such that

$$0 < \Delta X < \frac{1}{2}\varepsilon, \quad 0 < \Delta Y < \frac{1}{2}\varepsilon .$$

Partition the circumscribed rectangle into a grid of ΔX by ΔY subrectangles. (If ΔX does not evenly divide $a_2 - a_1$, the right column of subrectangles will be smaller, and similarly for ΔY.) Let G be the set of all subrectangles and

$$G_1 = \{E \in G: E \text{ is included in the interior of } D\},$$

$$G_2 = \{E \in G: E \text{ meets the boundary of } D\}.$$

Since $\Delta X + \Delta Y < \varepsilon$, we have $E \subseteq D_0$ whenever $E \in G_1 \cup G_2$. We wish to estimate the difference between B(D) and the finite Riemann sum $\Sigma\Sigma_D\, h(x,y)\Delta X\Delta Y$. Let $p(\Delta X,\Delta Y)$ be the total area of the subrectangles which meet the boundary of D, that is,

$$p(\Delta X,\Delta Y) = \text{area of } \bigcup G_2 .$$

For each $E \in G_1 \cup G_2$ let

$$H(E) = h(x,y)\Delta X\Delta Y$$

where (x,y) is the lower left corner of E, and let $q(\Delta X,\Delta Y)$ be the maximum of the values

$$\frac{|B(E) - H(E)|}{\Delta X \Delta Y}, \quad E \in G_1 \cup G_2 .$$

Since $0 \le h(x,y) \le M$ we have

$$0 \le H(E) \le M\Delta X\Delta Y,$$
$$0 \le B(E) \le (M + q(\Delta X,\Delta Y))\Delta X\Delta Y .$$

By Lemma 3,

$$\Sigma_{E \in G_2} B(D \cap E) \le \Sigma_{E \in G_2} B(E) .$$

It follows that

$$\begin{aligned} &|B(D) - \Sigma\Sigma_D\, h(x,y)\Delta X\Delta Y| \\ &\le |\Sigma_{E \in G_1} B(E) - H(E)| + \Sigma_{E \in G_2} B(D \cap E) + \Sigma_{E \in G_2} H(E) \\ &\le \Sigma_{E \in G_1} |B(E) - H(E)| + \Sigma_{E \in G_2} (B(E) + H(E)) \\ &\le (a_2 - a_1)(B_2 - B_1)q(\Delta X,\Delta Y) + p(\Delta X,\Delta Y)(2M + q(\Delta X,\Delta Y)). \end{aligned}$$

Now consider the given positive infinitesimals Δx, Δy. By the Solution Axiom,

$$\begin{aligned} &|B(D) - \Sigma\Sigma_D\, h(x,y)\Delta x\Delta y| \\ &\le (a_2 - a_1)(B_2 - B_1)q(\Delta x,\Delta y) + p(\Delta x,\Delta y)(2M + q(\Delta x,\Delta y)). \end{aligned}$$

By Lemma 4 and the Solution Axiom, $0 < p(\Delta x,\Delta y) < \varepsilon_0$ for every positive real ε_0, so $p(\Delta x,\Delta y) \approx 0$. By the Partial Solution Theorem there is a Δx by Δy element of area ΔD such that ΔD meets D and

$$\frac{|B(\Delta D) - h(x,y)\Delta x\Delta y|}{\Delta x\Delta y} = q(\Delta x,\Delta y) .$$

Therefore $\Delta D \subsetneq D_0^*$ and by hypothesis (iii), $q(\Delta x,\Delta y)$ is infinitesimal. We conclude that

$$|B(D) - \Sigma\Sigma_D\, h(x,y)\Delta x\Delta y| \approx 0,$$

whence

$$B(D) = \int\int_D h(x,y)\,dx\,dy . \qquad \dashv$$

In Elementary Calculus the Infinite Sum Theorem for two variables is used to justify integration formulas for the volume between two surfaces, for mass, first moments, and moments of inertia of plane objects, and for the area of a surface. Another application is given in the next section.

12D CHANGE OF VARIABLES IN DOUBLE INTEGRALS (§12.5)

In Elementary Calculus the Infinite Sum Theorem is used to prove the formula for a double integral over a region

$$D = \{(r,\theta): \alpha \le \theta \le \beta,\ a(\theta) \le r \le b(\theta)\}$$

in polar coordinates:

$$\iint_D f(x,y)\,dA = \int_\alpha^\beta \int_{a(\theta)}^{b(\theta)} f(r\cos\theta,\ r\sin\theta)r\,dr\,d\theta .$$

Here we shall use the Infinite Sum Theorem to obtain the general formula for a double integral with a change of variables. For comparison we first state the simpler result for single integrals (§4.7 in Elementary Calculus).

THEOREM 1

Suppose f is continuous on an open interval I, g maps an open interval J into I, and g is continuously differentiable. Then for all a,b in J,

$$\int_{g(a)}^{g(b)} f(u)\,du = \int_a^b f(g(t))g'(t)\,dt .$$

Proof Let F be an antiderivative of f on I and let $H(t) = F(g(t))$ for $t \in I$. By the Chain Rule,

$$H'(t) = f(g(t))g'(t),\ t \in I.$$

Then by the Fundamental Theorem of Calculus,

$$\int_{g(a)}^{g(b)} f(u)\,du = F(g(b)) - F(g(a)) = H(b) - H(a)$$

$$= \int_a^b f(g(t))g'(t)\,dt . \qquad \dashv$$

In the analogous theorem for two variables, the mapping g is replaced by a smooth transformation from an open region in the (s,t) plane into the (x,y) plane, and the derivative of g is replaced by the Jacobian of the transformation. Our elementary treatment of double integrals over basic closed regions is not adequate for integration by change of variables. The difficulty is that the image of a basic closed region under a smooth transformation is not necessarily a basic closed region. To give a natural treatment we need the classical theory of Jordan content. Once this theory is developed, the Infinite Sum Theorem can be used to obtain the change of variables formula. Our main purpose in this section is to show how the Infinite Sum Theorem is used. To prepare the way we sketch a hyperreal form of Jordan content.

Let D be a bounded set in the plane. The circumscribed rectangle E of D is the intersection of all closed rectangles containing D. Given positive real numbers Δx and Δy, partition E into a grid of Δx by Δy subrectangles. Let $\underline{C}(\Delta x,\Delta y)$ be the sum of the areas of the subrectangles entirely within D, and let $\overline{C}(\Delta x,\Delta y)$ be the sum of the areas of the subrectangles which meet D. Then $\underline{C}(dx,dy)$ and $\overline{C}(dx,dy)$ are defined for all positive infinitesimals dx and dy. We say that D has Jordan content A(D) if for all positive infinitesimals dx and dy, we have

$$A(D) = st(\underline{C}(dx,dy)) = st(\overline{C}(dx,dy)) .$$

Intuitively, the total area of the infinitesimal subrectangles included in D* has standard part A(D). Also, the total area of the infinitesimal subrectangles which meet D* has standard part A(D). The Jordan content of D is also called the area of D.

Let D be a set with Jordan content A(D), and f a continuous function on an open region containing the closure of D. The double Riemann sum and double integral are defined exactly as in the case of a basic closed region. Specifically, for positive infinitesimal dx and dy, we partition the circumscribed rectangle E of D into a dx by dy grid and define

$$\Sigma\Sigma_D\, f(x,y)\,dxdy = \sum_{(x_K,y_L)\in D^*}\sum f(x_K,y_L)\,dxdy ,$$

$$\iint_D f(x,y)\,dxdy = st(\Sigma\Sigma_D\, f(x,y)\,dxdy) .$$

The basic properties of double integrals are readily generalized to integrals over sets which have Jordan content, and take the following form.

INDEPENDENCE OF dx AND dy

The double integral $\iint_D f(x,y)dxdy$ does not depend on the infinitesimals dx and dy.

ADDITION PROPERTY

If $D = D_1 \cup D_2$ where $D_1 \cap D_2$ has Jordan content zero, and D_1, D_2 have Jordan content, then

$$\iint_D f(x,y)dxdy = \iint_{D_1} f(x,y)dxdy + \iint_{D_2} f(x,y)dxdy .$$

CYLINDER PROPERTY

Suppose $m \le f(x,y) \le M$ for $(x,y) \in D$. Then

$$m \cdot A(D) \le \iint_D f(x,y)dxdy \le M \cdot A(D) .$$

Taking $f(x,y) = 1$, we see at once that

$$A(D) = \iint_D 1 \, dx \, dy .$$

It follows that if D is a basic closed region, then D has Jordan content equal to the area of D as defined in Chapter 6.

SUM RULE

If f and g are continuous on D_0, then

$$\iint_D (f(x,y) + g(x,y))dxdy = \iint_D f(x,y)dxdy + \iint_D g(x,y)dxdy .$$

This follows from the analogous formula for double Riemann sums. We now introduce the notion of a smooth transformation.

DEFINITION

Let D_0 be an open set in the (s,t) plane. A <u>smooth transformation</u> T on D_0 is a mapping

$$T(s,t) = (x,y)$$

of D_0 into the (x,y) plane such that each of the functions $x = g(s,t)$ and $y = h(s,t)$ has continuous partial derivatives on D_0.

The image of a subset $D \subseteq D_0$ under T is the set

$$T(D) = \{T(s,t): (s,t) \in D\}.$$

The Jacobian of a smooth transformation T, denoted by $J(T)$ or $\frac{\partial(x,y)}{\partial(s,t)}$, is the function

$$\frac{\partial(x,y)}{\partial(s,t)} = \begin{vmatrix} \frac{\partial x}{\partial s} & \frac{\partial x}{\partial t} \\ \frac{\partial y}{\partial s} & \frac{\partial y}{\partial t} \end{vmatrix} = \frac{\partial x}{\partial s}\frac{\partial y}{\partial t} - \frac{\partial x}{\partial t}\frac{\partial y}{\partial s}.$$

Thus $\frac{\partial(x,y)}{\partial(s,t)}$ is a continuous function with domain D_0.

For example, the polar coordinate transformation

$$x = r\cos\theta,\ y = r\sin\theta$$

is smooth and its Jacobian is

$$\frac{\partial(x,y)}{\partial(r,\theta)} = \begin{vmatrix} \cos\theta & -r\sin\theta \\ \sin\theta & r\cos\theta \end{vmatrix} = r.$$

We shall use the following two lemmas which can be found in many advanced calculus texts (for example, the textbook of Buck).

LEMMA 1 Let T be a smooth transformation which is one to one and has nonzero Jacobian on an open set D_0. Let D be a closed bounded subset of D_0 which has Jordan content. Then

(i) $T(D_0)$ is an open set.
(ii) $T(D)$ is a closed bounded set which has Jordan content.
(iii) T maps the interior of D onto the interior of $T(D)$, and the boundary of D onto the boundary of $T(D)$.

LEMMA 2 Suppose T is a smooth transformation with nonzero Jacobian on an open set D_0, and the partial derivatives $\partial x/\partial s$, $\partial x/\partial t$, $\partial y/\partial s$, $\partial y/\partial t$ are bounded on D_0. Then for every real $\varepsilon > 0$ there is a real $\delta > 0$

such that for every square $\Delta D \subseteq D_0$ with side $\Delta s < \delta$, the area of the image of ΔD is within $\varepsilon \Delta s^2$ of the Jacobian times the area of ΔD, that is,

$$|A(T(\Delta D)) - |J(T)|\Delta s^2| < \varepsilon \Delta s^2 .$$

The number $|J(T)\Delta s^2|$ is the area of the parallelogram with vertex $T(s,t)$ and sides

$$\Delta s\left(\frac{\partial x}{\partial s}\vec{i} + \frac{\partial y}{\partial s}\vec{j}\right), \quad \Delta s\left(\frac{\partial x}{\partial t}\vec{i} + \frac{\partial y}{\partial t}\vec{j}\right).$$

The lemma is proved by showing that the boundaries of $T(\Delta D)$ are close to the boundaries of the parallelogram.

CHANGE OF VARIABLES THEOREM

Let T be a smooth transformation which is one to one and has nonzero Jacobian on an open set D_0. Then for every continuous function $f(x,y)$ on $T(D_0)$ and every basic closed region $D \subseteq D_0$,

$$\iint_{T(D)} f(x,y)\,dxdy = \iint_D f(T(s,t))\,|J(T)|\,dsdt .$$

For example, if D is a basic closed region

$$D = \{(r,\theta): \alpha \le \theta \le \beta,\ a(\theta) \le r \le b(\theta)\}$$

where $\alpha \le \beta < \alpha + 2\pi$ and $0 < a(\theta) \le b(\theta)$, then the polar coordinate transformation T satisfies the hypotheses of the theorem, and we obtain the formula

$$\iint_{T(D)} f(x,y)\,dxdy = \iint_D f(r\cos\theta, r\sin\theta)\,r\,dr\,d\theta.$$

We wrote $\beta < \alpha + 2\pi$ to make T one to one, and $0 < a(\theta)$ to make $r = J(T) \ne 0$ in D. But by passing to a limit we see that the polar integration formula is valid even for

$$\alpha \le \beta \le \alpha + 2\pi,\ 0 \le a(\theta) \le b(\theta) .$$

Proof of the Theorem Let D be a basic closed region contained in D_0. By Lemma 2 in the preceding section we may assume that the partial derivatives of x and y are bounded on D_0. We may also assume that $f(T(s,t))$ is bounded on D_0, so that $f(x,y)$ is bounded on $T(D_0)$. We first give the proof in the case that $f(x,y) \geq 0$ on $T(D_0)$. Let $B(D_1)$ be defined for basic closed regions $D_1 \subseteq D_0$ by

$$B(D_1) = \iint_{T(D_1)} f(x,y)\,dxdy .$$

By the Cylinder Property, $B(D_1) \geq 0$ for all D_1. If $D_1 = D_2 \cup D_3$ where D_2 and D_3 have disjoint interiors, then $T(D_1) = T(D_2) \cup T(D_3)$ and by Lemma 1 in this section, $T(D_2)$ and $T(D_3)$ have disjoint interiors. Thus by the Addition Property for integrals, $B(D_1) = B(D_2) + B(D_3)$. Let Δs be positive infinitesimal and let $\Delta t = \Delta s$. By Lemma 2 in this section and the Solution Axiom, we have

$$A(T(\Delta D)) \approx |J(T)|\Delta s^2 \quad (\text{compared to } \Delta s^2) ,$$

and

$$\frac{A(T(\Delta D))}{\Delta s^2} \approx |J(T)| .$$

By the Hyperreal Extreme Value Theorem, $f(x,y)$ has a minimum value m and maximum value M on $T(\Delta D)$. Since T is continuous, every point of $T(\Delta D)$ is infinitely close to $T(s,t)$ where (s,t) is the lower left corner of ΔD. By continuity of f, m and M are infinitely close to $f(T(s,t))$. Then by the Cylinder Property and Solution Axiom,

$$mA(T(\Delta D)) \leq B(\Delta D) \leq MA(T(\Delta D)),$$

whence

$$B(\Delta D) \approx f(T(s,t))A(T(\Delta D)) \quad (\text{compared to } A(T(\Delta D))) .$$

Therefore

$$\frac{B(\Delta D)}{\Delta s^2} = \frac{B(\Delta D)}{A(T(\Delta D))}\,\frac{A(T(\Delta D))}{\Delta s^2} \approx f(T(s,t))|J(T)| ,$$

and

$$B(\Delta D) \approx f(T(s,t))|J(T)|\Delta s^2 \quad (\text{compared to } \Delta s^2).$$

It follows from the Infinite Sum Theorem that

$$B(D) = \iint_D f(T(s,t))\,|J(T)|\,ds\,dt \ .$$

Now we consider the general case where f is not necessarily ≥ 0 in D_0. Since f is bounded on D_0 there is a real number $c > 0$ such that $f(x,y) \geq -c$ on D_0. Then $f(x,y) + c \geq 0$ on D_0, so by the previous result,

$$\iint_{T(D)} (f(x,y) + c)\,dx\,dy = \iint_D (f(T(s,t)) + c)\,|J(T)|\,ds\,dt \ ,$$

$$\iint_{T(D)} c\ dx\ dy = \iint_D c\,|J(T)|\,ds\,dt \ .$$

Using the Sum Rule for integrals,

$$\begin{aligned}
&\iint_{T(D)} f(x,y)\,dx\,dy \\
&\quad = \iint_{T(D)} (f(x,y) + c)\,dx\,dy - \iint_{T(D)} c\ dx\,dy \\
&\quad = \iint_D (f(T(s,t)) + c)\,|J(T)|\,ds\,dt - \iint_D c\,|J(T)|\,ds\,dt \\
&\quad = \iint_D f(T(s,t))\,|J(T)|\,ds\,dt \ .
\end{aligned}$$

⊣

INDEX

NOTE: A CONFUSING MISPRINT

The first printing of Elementary Calculus has the following misprint. Page 33, line 9 reads

Every real number is infinite.

This should be changed to

Every real number is finite.

Be sure to tell your students to make this correction in their copies of Elementary Calculus.